C000002092

Series FM1 no. 34

Birth Statistics

Review of the Registrar General on births and patterns of family building in England and Wales, 2005

(Including supplement: Conception Statistics Conceptions for women resident in England and Wales, 2004)

Contact points
For enquiries about this publication, contact
Vital Statistics Outputs Branch
Tel: 01329 813758
E-mail: vsob@ons.gsi.gov.uk

For general enquiries, contact the National Statistics
Customer Contact Centre on: 0845 601 3034
(minicom: 01633 812399)
E-mail: info@statistics.gsi.gov.uk
Fax: 01633 652747
Post: Room 1015, Government Buildings,
Cardiff Road, Newport NP10 8XG

You can also find National Statistics on the Internet at:
www.statistics.gov.uk

Palgrave hardcopy:
ISBN (10) 1-40399-918-X
ISBN (13) 978-1-4039-9918-4

About the Office for National Statistics
The Office for National Statistics (ONS) is the government
agency responsible for compiling, analysing and disseminating
many of the United Kingdom's economic, social and
demographic statistics, including the retail prices index, trade
figures and labour market data, as well as the periodic census
of the population and health statistics. It is also the agency that
administers the statutory registration of births, marriages and
deaths in England and Wales. The Director of ONS is also the
National Statistician and the Registrar General for England and
Wales.

A National Statistics publication
National Statistics are produced to high professional standards
set out in the National Statistics Code of Practice. They undergo
regular quality assurance reviews to ensure that they meet
customer needs. They are produced free from any political
influence.

Contents

List of main tables and appendices

3 Parents' age

4 Previous live-born children

5 Marriage duration

6 Multiple births

7 Area of usual residence

11 Socio-economic classification

Appendix: base populations

Annexes

1 Introduction

Birth statistics 2005 presents statistics on births occurring annually in England and Wales between 1995 and 2005.

The annual update on fertility trends in 2005 was published in *Population Trends* 126 in December 2006. Provisional data for 2005 were first published in *Population Trends* 124.

This volume is produced by the Office for National Statistics (ONS). It is published under the National Statistics logo, the designation guaranteeing that those outputs have been produced to high professional standards set out in a Code of Practice, and have been produced free from political influence.

The registration of life events (births, deaths and marriages) is a service carried out by the Local Registration Service in partnership with the General Register Office (GRO) in Southport, which is part of ONS. ONS was formed on 1 April 1996, bringing together the Office of Population Censuses and Surveys (OPCS) and the Central Statistical Office. OPCS is referred to in this volume for historic events and publications.

1.1 Tables in this volume

The tables presented in this volume are set out in 11 sections, covering various fertility topics including characteristics of the birth and of the parents. Tables in **Section 10** analyse trends in cohort fertility; otherwise, the tables use period fertility measures.

For brevity, the time series shown here have been limited to a run of 11 years at most. Figures for earlier years are shown in earlier volumes of *Birth statistics*. This series includes a volume of historical fertility statistics[1] containing some time series back to 1837, the year of the introduction of compulsory birth registration. More detailed time series are also included for years from 1938, the year in which the first Population Statistics Act came into force - see section 2.14.

As described in *Birth statistics* 2002, conception figures are now published separately in a supplementary volume. The supplementary volume *Conception statistics* 2004 was published in December 2006.

Table 1.5 shows the number of marriages and marriage rates for the years 1994 to 2004.

Tables 9.1 and **9.6** present live births by birthplace of mother. Data for previous years have been compiled to be consistent with the country classification listing as constituted in 2005.

Table 9.5 shows the total fertility rates by mother's country of birth and uses population denominators from the Censuses of 1991 and 2001. It aggregates countries into groupings that provide denominators large enough for reliable rates. Surveys conducted between Censuses are unable to provide reliable denominators for intervening years.

Data from 2001 onwards have been coded to the new socio-economic classification as defined by occupation and employment status. Therefore **Tables 11.1** to **11.5** have been adjusted to show data for the five years 2001, 2002, 2003, 2004 and 2005 only. See section 3.10 for further information.

1.2 Data analysed in this volume

The data used in this volume are summarised on the following pages. The items covered are collected for both live births and stillbirths.

The information used in tables in this volume is based largely on the details collected when births are registered. Most of the information, for both live births and stillbirths, is supplied to registrars by one or both parents. For live births, details of **birthweight** are notified to the local health authority by the hospital where the birth took place, or by the midwife or doctor in attendance at the birth. These details are then supplied to the registrar.

For stillbirths, details of **cause of death, duration of pregnancy and weight of foetus** are supplied on a certificate or notification by the doctor or midwife either present at the birth, or who examined the body. The certificate or notification is then taken by the informant to a registrar. The registrar will use all this information to complete a draft entry *Form 309* (**Annex A**) for a live birth or *Form 308* (**Annex B**) for a stillbirth.

The **date of birth** is supplied in a conventional way, except that where more than one child is liveborn at a confinement, then time is also recorded. **Place of birth** is entered as the usual name and the address of a hospital, maternity home or other communal establishment, or the address of a private dwelling. ONS then codes place of birth to one of the groups of places in **Table 8.1**. The **sex** of the child is also recorded.

Although the **birthplace of the parents** may be recorded in detail if this was in the United Kingdom, the main interest in this volume is with parents born outside the United Kingdom - see **Tables 9.1** to **9.6**.

Table number	Year(s)	Area	Numbers/ rates/ percent- ages (N/R/P)	Live births/ stillbirths (LB/SB)	Mater- nities/ pater- nities (M/P)	Age of mother (Year of birth)	Age of father	Period of occur- rence	Period of regist- ration	Sole/ joint registra- tion (S/J)	Within/ outside marriage (W/O)
	Summary										
1.1	1995-2005	E&W	NR	LB							WO
1.2	1995-2005	E&W	NR	SB							WO
1.3	1995-2005	E&W	NR	LB							
1.4	1995-2005	E&W	R	LB							
1.5	1994-2004	E&W	NR								
1.6	1995-2005	E&W	Mean	LB		*					WO
1.7	1995-2005	E&W	Mean	LB		*					
1.8	1994-2004	E&W	P	LB/SB		*					W
1.9	1995-2005	E&W	R	LB		*					WO
	Seasonality										
2.1	1995-2005	E&W	NR	LB					*		
2.2	1995-2005	E&W	N	SB					*		
2.3	1995-2005	E&W	NR	LB					*		
2.4	2005	E&W	N	LB/SB	M				*		WO
2.5	2005	E&W	N	LB					*	*	
	Parents' age										
3.1	1995-2005	E&W	NR	LB		*					WO
3.2	2005	E&W	N	LB/SB	M	*					WO
3.3	1995-2005	E&W	NR	LB			*				W
3.4	2005	E&W	N	LB/SB	P		*			J	WO
3.5	2005	E&W	R	LB/SB	P		*				W
3.6	2005	E&W	N	LB		*	*				W
3.7	2005	E&W	N	SB		*	*				W
3.8	2005	E&W	N	LB		*	*			SJ	O
3.9	1995-2005	E&W	NP	LB		*				SJ	O
3.10	1995-2005	E&W	NP	LB		*				J	O
	Previous children										
4.1	1995-2005	E&W	N	LB		*					W
4.2	2005	E&W	N	LB		*					W
4.3	2005	E&W	NR	LB/SB		*					W
	Marriage duration										
5.1	1995-2005	E&W	N	LB		*					W
5.2	1995-2005	E&W	N	LB		*					W
5.3	2005	E&W	N	LB		*					W
	Multiple births										
6.1	1995-2005	E&W	NR		M	*					WO
6.2	2005	E&W	N	LB/SB	M	*					
6.3	2005	E&W	NR		M	*					W
6.4	2005	E&W	N	LB/SB	M	*					
	Area of usual residence										
7.1	2005	E&W+	NR	LB							WO
7.2	2005	E&W+	NR	LB							WO
7.3	2005	E&W+	NR	LB		*					WO
7.4	2005	E&W+	NR	LB		*					
7.5	2005	E&W+	NP	LB							
7.6	2005	E&W+	NR	SB							WO
7.7	2005	E&W+	NP	SB							

Sex	Previous live-born children	Birth-weight	Place of confine-ment	Country of birth mother/ father (M/F)	Socio-economic classification of father/ husband	Usual residence/ place of occurrence (R/O)	Marriage order	Year(s)	Table number
								Summary	
*								1995-2005	1.1
*								1995-2005	1.2
								1995-2005	1.3
								1995-2005	1.4
								1994-2004	1.5
	*							1995-2005	1.6
	*							1995-2005	1.7
						*		1994-2004	1.8
	*							1995-2005	1.9
								Seasonality	
								1995-2005	2.1
								1995-2005	2.2
								1995-2005	2.3
*								2005	2.4
								2005	2.5
								Parents' age	
								1995-2005	3.1
*								2005	3.2
								1995-2005	3.3
*								2005	3.4
*								2005	3.5
								2005	3.6
								2005	3.7
								2005	3.8
								1995-2005	3.9
								1995-2005	3.10
								Previous children	
	*							1995-2005	4.1
	*							2005	4.2
*	*							2005	4.3
								Marriage duration	
								1995-2005	5.1
						*		1995-2005	5.2
						*		2005	5.3
								Multiple births	
								1995-2005	6.1
*								2005	6.2
	*							2005	6.3
*								2005	6.4
								Area of usual residence	
						R		2005	7.1
						R		2005	7.2
						R		2005	7.3
						R		2005	7.4
		*				R		2005	7.5
						R		2005	7.6
		*				R		2005	7.7

Table number	Year(s)	Area	Numbers/ rates/ percent-ages (N/R/P)	Live births/ stillbirths (LB/SB)	Mater-nities/ pater-nities (M/P)	Age of mother (Year of birth)	Age of father	Period of occur-rence	Period of regist-ration	Sole/ joint registra-tion (S/J)	Within/ outside marriage (W/O)
	Place of confinement										
8.1	2005	E&W	N		M	*					WO
8.2	2005	E&W+	N		M						
8.3	2005	E&W+	NR	LB/SB	M						
	Parents' birthplace										
9.1	1995, 2000-2005	E&W	NP	LB							
9.2	2005	E&W+	NP	LB							
9.3	2005	E&W	N	LB							
9.4	2005	E&W	N	LB		*					
9.5	1991 and 2001	E&W	R	LB							
9.6	1995, 2002-2005	E&W	NP	LB							WO
	Cohorts										
10.1	1920-1990	E&W	R	LB		*					
10.2	1920-1990	E&W	R	LB		*					
10.3	1920-1990	E&W	R	LB		*					
10.4	1920-1986	E&W	R	LB		*		*			WO
10.5	1920-1985	E&W	P	LB		*					
	Socio-economic classification										
11.1	2001-2005	E&W	N	LB		*					W
11.2	2001-2005	E&W	NP	LB		*					W
11.3	2001-2005	UK and E&W	Median	LB							W
11.4	2001-2005	E&W	Mean	LB		*		*			W
11.5	2001-2005	E&W	N	LB		*				J	O

The **mother's usual address** is entered, as is that of the informant where appropriate. This information is used for tables showing usual residence of mother **7.1** to **7.7**, as well as **Table 3.10**, which analyses jointly registered births outside marriage and whether the parents resided at the same address.

Occupation is recorded for the mother and for the father, if his name is entered in the register. The informant is asked whether the father/mother was in gainful employment at any time before the child's birth, and a description of the occupation may be recorded. The informant may not wish to have details of the father/mother's occupation entered in the register, but it may still be recorded for use in statistical analyses. If the father is unemployed, his last full-time occupation will be recorded. As discussed in section 3.10, this information is used for analyses of socio-economic classification as defined by occupation in **Tables 11.1** to **11.5.**

Informants are also required to provide further information, treated as confidential, under the provisions of the Population Statistics Acts (PSA), as below:

(i) the **father's date of birth,** if his name is entered in the register;

(ii) the **mother's date of birth;**

if the child's parents were married to each other at the time of the birth:

(iii) the **date of the parents' marriage;**

(iv) **whether the mother has been married more than once;**

(v) **number of previous children** by her present husband and any former husband, (a) **born alive,** and (b) **stillborn.**

These confidential details are used extensively in this volume, in particular for analyses of **age of mother and father,** of **marriage duration,** and of **birth order.** See section 1.3 below for details of issues affecting the quality of these variables in 2005.

Other statistical information collected at registration includes the economic activity of the parents, i.e. **industry** and **employment status,** and whether the confinement resulted in a **multiple birth.**

1.3 Issues affecting the quality of the data in this volume

In this volume figures for 2004 are those published on 30 August 2006.[2]

Sex	Previous live-born children	Birth-weight	Place of confine-ment	Country of birth mother/ father (M/F)	Socio-economic classification of father/ husband	Usual residence/ place of occurrence (R/O)	Marriage order	Year(s)	Table number
								Place of confinement	
	*		*					2005	8.1
			*			RO		2005	8.2
						O		2005	8.3
								Parents' birthplace	
				M				1995, 2000-2005	9.1
				M		R		2005	9.2
				MF				2005	9.3
				M				2005	9.4
				M				1991 and 2001	9.5
	*			M				1995, 2002-2005	9.6
								Cohorts	
								1920-1990	10.1
								1920-1990	10.2
								1920-1990	10.3
	*							1920-1986	10.4
	*							1920-1985	10.5
								Socio-economic classification	
	*				*			2001-2005	11.1
					*			2001-2005	11.2
					*			2001-2005	11.3
	*				*			2001-2005	11.4
					*			2001-2005	11.5

Under the Population Statistics Act (PSA) certain data items are collected at the registration of a birth (mother's date of birth, father's date of birth (where his name is on the certificate), and for births within marriage the date of marriage and numbers of previous live born and stillborn children). If any of these data items is missing, an appropriate value is imputed.

In 2005, the percentage of records without mother's date of birth details, rose from 0.8 per cent in 2004 to 1.1 per cent in 2005. Although this affected only a small number of records, they were clustered in particular registration districts.

Since 2004 all missing values for PSA data items have been re-imputed using CANCEIS (Canadian Census Edit and Imputation System)[3]. In 2005, this has improved the distribution for mother's age, especially in small areas, and for the other PSA data items, compared with the previous imputation system.

In 1999 the proportion of live birth registrations without Population Statistics information received from one Register Office was higher than usual due to a combination of circumstances. The missing data on those records were imputed using a random sample of data from the particular area from the previous three years. This was a change from the usual method, but was used to improve the quality of the imputations. Procedures were put in place which mean that such a problem is unlikely to recur. For further information about this see section C.2 of the 1999 volume.[4]

After the publication of the 1997 volume in this series, an error was discovered in the births database that resulted in 1,002 live births being excluded from the published 1997 statistics. Details of the error were included in section D.1 of the 1998 volume.[5] All time series tables in this and other volumes since 1997 reflect the corrected data for 1997.

Since the dataset for the 2000 volume, a small number of very late registrations have been excluded each year from the official statistics. Inclusion of these very late registrations in the statistical dataset was found to have an adverse effect on the quality of infant mortality data when linked with the live birth data. The annual dataset now includes only those births occurring in the reference year, and late registrations of births occurring in the year previous to the reference year. See section 2.2 for further discussion of occurrences and registrations. The numbers of late registrations included in, and numbers of very late registrations excluded from the statistics are shown in **Table A**.

1.4 Associated publications and the National Statistics website

The National Statistics website (**www.statistics.gov.uk**) provides a comprehensive source of freely available vital statistics and ONS products. More information on the National Statistics website can be obtained from the contact address in section 2.16.

Historic data, and figures for UK countries

Comparable statistics for earlier years, and separate statistics for Scotland and Northern Ireland are published as follows:

England and Wales: from 1974 - 2005 in FM1 *Birth statistics;* for earlier years in the *Registrar General's Statistical Review of England and Wales.* Data for the years 1837-1983 are summarised in an earlier volume in the FM1 series.[1]

Scotland: in the *Annual Report of the Registrar General for Scotland.* See website: **www.scotland.gov.uk/Topics/Statistics**

Northern Ireland: in the *Annual Report of the Registrar General for Northern Ireland.* See website: **www.nisra.gov.uk**

A summary of fertility statistics for the United Kingdom and constituent countries appears in the *Annual Abstract of Statistics,* issued by ONS. Similar data also appears in the ONS quarterly journals *Population Trends* and *Health Statistics Quarterly.* Data for Europe are published in the Council of Europe annual volume *Recent demographic developments in Europe,* and the Eurostat publication *Population Statistics.* See website: **www.eustatistics.gov.uk.** Statistics for United Nations member countries appear in the annual *UN Demographic Yearbook.*

Other related annual reference volumes published by ONS include:

- **DH3** *(Mortality statistics: childhood, infant and perinatal),* which contains data on stillbirths, infant deaths, and childhood deaths. It includes figures for infant deaths linked to their corresponding birth records, as well as birthweight data for health authorities analysed separately by age, socio-economic classification and parity.

- **FM2** *(Marriage, divorce and adoption statistics),* which provides data on marriages by age of bride and bridegroom, on divorces by ages of children involved, and on adoptions by age of child at adoption.

Fertility data are also published annually in the ONS volume *Key population and vital statistics.*[6] This volume provides data on population, births, deaths, and migration, for administrative areas in the United Kingdom, including local and health authorities/boards.

Population Trends and Health Statistics Quarterly publications

Up to 1998 ONS published annual data in Monitors, known as the series FM1 for conceptions and live births. These contained basic information on annual conceptions and live birth registrations, and were issued soon after the data became available. However, these publications have been discontinued and since 1999 these data have appeared in Reports issued in the quarterly journal *Population Trends.* Since the beginning of 1999, ONS has published two quarterly journals: *Population Trends,* which now has an emphasis on population and demography, covering most fertility topics, and *Health Statistics Quarterly,* covering mortality and health topics, including abortions, and some other fertility data. The annual Report on live births by local and health authority areas/boards is published in the Summer issue of *Population Trends.*

Population Trends and *Health Statistics Quarterly* both contain regular quarterly reference tables on a variety of population and health topics; for fertility these include analyses of births (numbers and rates) by age of mother.

Table A Late registrations included in, and excluded from, the births annual datasets

Annual dataset year	Number of late registrations from the previous year included in the dataset	Number of very late registrations excluded from the dataset
2000	519	34
2001	195	20
2002	161	17
2003	207	18
2004	320	9
2005	307	21

1.5 Other publications

Some other recent background information on fertility data and other relevant articles and publications are listed below. Most are from the journal *Population Trends*, but copies of any not easily available may be obtained from ONS - see section 2.16.

Trends in fertility

- Chamberlain, J and Gill, B (2005). Chapter 5: Fertility and Mortality, *Focus on People and Migration* Palgrave Macmillan: Basingstoke, 71-90.
- ONS. Annual update: Births in 2005 in England and Wales, *Population Trends* 126, pp 64-68, Winter 2006.
- ONS. Annual update: Births in 2004 in England and Wales, *Population Trends* 122, pp 87-91, Winter 2005.
- Berrington, A. Perpetual postponers? Women's, men's and couple's fertility intentions and subsequent fertility behaviour. *Population Trends* 117, pp 9-19, Autumn 2004.
- Smallwood, S. Characteristics of sole registered births and the mothers who register them. *Population Trends* 117, pp 20-26, Autumn 2004.
- ONS. Annual update: Births in 2003 in England and Wales. *Population Trends* 118, pp 72-76, Winter 2004.
- Kiernan, K and Smith, K. Unmarried parenthood: new insights from the Millennium Cohort Study. *Population Trends* 114, pp 26-33, Winter 2003.
- Murphy, M and Grundy, E. Mothers with living children and children with living mothers: the role of fertility and mortality in the period 1911-2050. *Population Trends* 112, pp 36-44, Summer 2003.
- Smallwood, S. The effects of changes in timing of childbearing on measuring fertility in England and Wales. *Population Trends* 109, pp 36-45, Autumn 2002.
- Joshi, H and Smith, K. The millennium cohort study. *Population Trends* 107, pp 30-34, Spring 2002.
- Botting, B and Dunnell, K. Trends in fertility and contraception in the last quarter of the 20th century. *Population Trends* 100, pp 32-40, Summer 2000.
- Armitage, R and Babb, P. Population Review (4): Trends in fertility. *Population Trends* 84, pp 7-13, Summer 1996.

Age patterns

- Rendall, M. How important are inter-generational cycles of teenage motherhood in England and Wales? A comparison with France. *Population Trends* 111, pp. 27-37, Spring 2003.
- Babb, P. Fertility of the over forties. *Population Trends* 79, pp 34-36, Spring 1995.
- Babb, P. Teenage conceptions and fertility in England and Wales, 1971-91. *Population Trends* 74, pp 12-17, Winter 1993.

Socio-economic classification

- Donkin, A, Lee, Y and Toson, B. Implications of changes in the UK social and occupational classifications in 2001 for vital statistics. *Population Trends* 107, pp 23-29, Spring 2002.

Birth order

- Rendall, M, Couet, C, Lappegard, T, Robert-Bobée, I, Rønsen, M and Smallwood, S. First births by age and education in Britain, France and Norway. *Population Trends* 121, pp 27-34, Autumn 2005.
- Rendall, M and Smallwood, S. Higher qualifications, first-birth timing and further childbearing in England and Wales. *Population Trends* 111, pp 18-26, Spring 2003.
- Smallwood, S. New estimates of trends in birth order in England and Wales. *Population Trends* 108, pp 32-48, Summer 2002.
- Wood, R. Trends in multiple births 1938-1995. *Population Trends* 87, pp 29-35, Spring 1997.

Future levels of fertility

- Smallwood, S and Chamberlain, J. Replacement fertility, what has it been and what does it mean? *Population Trends* 119, pp 116-27, Spring 2005.
- Shaw, C. Interim 2003-based national population projections for the United Kingdom and constituent countries. *Population Trends* 118, pp 6-17, Winter 2004.
- Shaw, C. 2002-based national population projections for the United Kingdom and constituent countries. *Population Trends* 115, pp 6-15, Spring 2004.
- Smallwood, S. Fertility assumptions for the 2002-based national population projections. *Population Trends* 114, pp 8-18, Winter 2003.
- Shaw, C. Interim 2001-based national population projections for the United Kingdom and constituent countries. *Population Trends* 111, pp 7-17, Spring 2003.
- GAD. National population projections 2000–based (Series PP2 no 23). Fertility, Chapter 6, pp 18-21, The Stationery Office, 2002.
- Shaw, C. 2000-based national population projections for the United Kingdom and its constituent countries. *Population Trends* 107, pp 5-13, Spring 2002.
- Shaw, C. Assumptions for the 2000-based National Population Projections. *Population Trends* 105 pp 45-47, Autumn 2001.
- GAD. 1998-based national population projections (Series PP2 no 22). Fertility, Chapter 7, pp 23-26, The Stationery Office, 2000.
- Shaw, C. 1996-based national population projections for the United Kingdom and constituent countries. *Population Trends* 91, pp 43-49, Spring 1998.

- Craig, J. Replacement level fertility and future population growth. *Population Trends* 78, pp 20-22, Winter 1994.

Other

- Population and Demography Division, ONS. The UK population at the start of the 21st century. *Population Trends* 122, pp 7-17, Winter 2005.
- Hancock, R, Stuchbury, R and Tomassini, C. Changes in the distribution of marital age differences in England and Wales, 1963 to 1998. *Population Trends* 114, pp 19-25, Winter 2003.
- Smith, J, Chappell, R, Whitworth, A and Duncan, C. Implications of 2001 Census for local authority district mid-year population estimates. *Population Trends* 113, pp 20-31, Autumn 2003.
- Smallwood, S and Jefferies, J. Family building intentions in England and Wales: trends, outcomes and interpretations. *Population Trends* 112, pp 15-28, Summer 2003.
- Hindess, G. Population review of England and Wales, 2001. *Population Trends* 112, pp 7-14, Summer 2003.
- Ghee, C. Population review of 2000, England and Wales. *Population Trends* 106, pp 7-14, Winter 2001.
- Berthoud, R. Teenage births to ethnic minority women. *Population Trends* 104, pp 12-18, Summer 2001.
- Haskey, J. Having a birth outside marriage: the proportion of lone mothers and cohabiting mothers who subsequently marry. *Population Trends* 97, pp 6-18, Autumn 1999.
- Botting, B, Rosato, M and Wood, R. Teenage mothers and the health of their children. *Population Trends* 93, pp 19-28, Autumn 1998.
- Filakti, H. Trends in abortion 1990-1995. *Population Trends* 87, pp 11-19, Spring 1997.
- Armitage, R. Variation in fertility between different types of local area. *Population Trends* 87, pp 20-28, Spring 1997.

2 Notes and definitions

2.1 Base populations

The population figures in **Appendix Table 1**, which are used to calculate fertility rates in this volume, are mid-year estimates of the resident population of England and Wales based on the 2001 Census of Population. These estimates include members of HM and non-UK armed forces stationed in England and Wales, but exclude those stationed outside. ONS mid-year population estimates are updated figures using the most recent Census, allowing for births, deaths, net migration and ageing of the population.

In this volume, the population estimates used for the calculation of fertility rates are the latest consistent estimates available at the time of its production and were published as follows:

Population estimates by age and sex:
- 2005 estimates - published on 24 August 2006;

Population estimates by marital status:
- 2005 marital status estimates - published on 30 November 2006

Further details about population estimates can be found at the National Statistics website (www.statistics.gov.uk/popest).

2.2 Occurrences and registrations

Between 1994 and 2000, the cut-off date for inclusion in the annual dataset was births occurring in the reference year registered by 11 February of the following year, this being 42 days after 31 December, the legal time limit for registering a birth. For 2001, the cut-off date was extended to 25 February 2002 to allow increased capture of births registered late. This change means that the annual statistics are prepared on as close to a true occurrences basis as possible, which provides a purer denominator for calculating infant mortality rates.

To avoid artificially inflating the 2001 dataset through the increased capture of late registrations, the start date for the carry over of late registrations from births occurring in 2000 was similarly moved by two weeks.

The total number of births recorded in this 2005 volume includes:

a) births occurring in 2005 which were registered by 25 February 2006, and
b) births occurring in 2004 which were registered between 26 February 2005 and 25 February 2006, that is births in the previous year which had not been tabulated previously.

The number of births registered late in (b) was 307 in the 2005 dataset, compared with 519 in the 2000 dataset, reflecting the extended cut-off date.

In the 2000 volume the total number of births included:[7]

a) births which occurred in 2000 registered by 11 February 2001, and
b) births occurring in 1999 which were registered between 12 February 2000 and 11 February 2001, that is births in the previous year which had not been tabulated previously.

Total annual births for 1994 to 1999 were derived in a similar way, except that births for all earlier years were included in the annual totals, not just births in the previous year - see section 1.3 for more details. Up to 1993 the cut-off date was 31 January of the following year, but from 1994 this was then extended to the legal time limit by which a birth should be registered (42 days).

The number of births registered late in (b) was about 1,500 to 3,000 per year from 1987 to 1993, but from 1994 it fell to about 500 to 600 annually.

2.3 Areal coverage

The births recorded in this volume are those occurring (and then registered) in England and Wales. No distinction is made between births to civilians and births to non-civilians.

A birth to a mother whose usual residence is outside England and Wales is assigned to the country of residence. These births are included in total figures for England and Wales, but excluded from any sub-division of England and Wales. They are identified as a separate group in **Tables 7.1 to 7.7**.

2.4 Registration of births

Every registrar of births and deaths is required to secure the prompt registration of births occurring within the sub-district covered. Registration of a birth is legally required within 42 days of its occurrence, and the registrar will, if necessary, send a requisition to the person whose duty it is to register the birth.

Under the National Health Service Act 1977, births must also be notified, within 36 hours, to the Director of Public Health in the health authority where the birth occurred. This is carried out by the hospital where the birth took place, or by the midwife or doctor in attendance at the birth. Each month, a list of the births which have occurred in the sub-district is supplied to the registrar, who will then check whether every birth has been registered.

The following people are qualified to give information to the registrar concerning a birth:

a. the mother of the child, and the father if the child was born within marriage;
b. the occupier of the house in which the child was, to the knowledge of that occupier, born;
c. any person present at the birth;
d. any person having charge of the child.

The duty of giving information is placed primarily upon the parents of the child but, in the case of death or inability of the parents, the duty falls on one of the other qualified informants.

The particulars to be registered concerning a birth are prescribed by the Births and Deaths Registration Act 1953, and are covered in section 1.2. Certain other particulars are collected for statistical purposes under the Population Statistics Acts 1938 and 1960, and are not entered in the register. All details are entered on a draft entry *Form 309* (**Annex A**) for a live birth, or *Form 308* (**Annex B**) for a stillbirth. These are checked by the informant before being entered in the register.

The procedures and information required for stillbirths are similar to those for live births. The main difference is the recording of the cause of death of the stillborn child, on evidence given by the doctor or midwife present at the birth, or who examined the body.

Usually, information for the registration of a birth must be given personally by the informant to the registrar for the sub-district in which the birth occurred. However, since April 1997 an informant may supply this information to any registrar by making a declaration of these particulars. The declaration is sent to the registrar of the sub-district where the birth occurred, and that registrar will enter the particulars in the register.

2.5 Visitors and overseas registrations

As noted above, the coverage of this volume is of births occurring, and then registered, in England and Wales. Births to residents of England and Wales which are registered elsewhere are thus excluded, while births registered in England and Wales to mothers whose usual residence is elsewhere, are included. In 2005, there were 214 live births in England and Wales to visitors whose normal residence was outside England and Wales.

In 2005, 9,758 births occurring outside the United Kingdom to British nationals were voluntarily registered with British Consulates, British High Commissioners, or HM Armed Forces registration centres. Most of these, however, were births to women who had emigrated from the United Kingdom - that is, had lived outside the UK for at least one year - and were thus not residents of England and Wales. Such persons are not included in population estimates for England and Wales.

At any one time some women of childbearing age (defined as age 15-44), usually resident in England and Wales, are temporarily absent overseas. But most of these women are absent for only a short period, and it is unlikely that more than a few hundred per year give birth while overseas. Also, the number of births during 2005 to residents of England and Wales that were registered in Scotland and Northern Ireland were 177 and 43 respectively.

Thus, the number of births to residents of England and Wales occurring outside the country is likely to be of the same order as the number of births occurring in England and Wales to visitors resident elsewhere. The effect on fertility rates of the difference between the definitions used for birth event numerators and population denominators is assumed to be negligible.

2.6 Foundlings

Few, if any, details are known about abandoned children, and they are not included in the statistics given in this volume. However, these infants are included in the Abandoned Children Register maintained at the GRO in Southport. Four such entries were made in 2005.

2.7 Country of birth of parents

The country of birth of parents for children born in England and Wales has been recorded at birth registration since April 1969.

Country of birth groupings represent the Commonwealth and European Union as constituted in 2005. The details for country of birth groupings are shown in **Table B**. Further summary tables containing data on live births by parents' birthplace can be found on the website: **www.statistics.gov.uk/downloads/theme_population/ countryofbirth.**

Table B   Country groupings for birthplace of parents

United Kingdom	England, Wales, Scotland, Northern Ireland, Northern Ireland, Channel Islands, Isle of Man, UK (not otherwise stated)
Outside United Kingdom	
Irish Republic	Eire, Ireland (not otherwise stated)
Other European Union	Austria, Belgium, Cyprus, Czech Republic, Denmark, Estonia, Finland, France, Germany, Greece, Hungary, Italy, Latvia, Lithuania, Luxembourg, Malta, The Netherlands, Poland, Portugal, Slovakia, Slovenia, Spain, Sweden
Rest of Europe	All other European countries, including Turkey, Russia and the rest of the former Soviet republics
Commonwealth	
Australia, Canada and New Zealand	
New Commonwealth	
Bangladesh, India, Pakistan	
East Africa	Kenya, Malawi, Tanzania, Uganda, Zambia
Southern Africa	Botswana, Lesotho, Namibia, South Africa, Bantu Homelands, Bophuthatswana, Transkei, Venda, Walvis Bay, Swaziland
Rest of Africa	Cameroon, The Gambia, Ghana, Mauritius, Mozambique, Nigeria, Seychelles, Sierra Leone
Caribbean	Anguilla, Antigua and Barbuda, Bahamas, Barbados, Belize, Bermuda, British Virgin Islands, Cayman Islands, Dominica, Grenada, Guyana, Jamaica, Leeward Islands, Montserrat, St Christopher (St Kitts) and Nevis, St Lucia, St Vincent and the Grenadines, Trinidad and Tobago, Turks and Caicos Islands, West Indies, New Commonwealth, Other Caribbean Islands
Far East	Brunei, Malaysia, Singapore
Rest of New Commonwealth	Australian Antarctic Territory, Christmas Islands, Cocos (Keeling) Islands, Coral Sea Islands Territory, Heard and McDonald Islands, Norfolk Islands, British Antarctic Territory, British Indian Ocean Territory, Cook Islands, Falkland Islands, East Falkland, West Falkland, Fiji, Gibraltar, Kiribati, The Maldives, Nauru, New Hebrides, Niue, Papua New Guinea, Pitcairn Islands Group, St Helena and Dependencies, Ascension Island, Gouch Island, Inaccessible Island, Middle Island, Nightingale Island, Stoltenhoff Island, Tristan Da Cunha, Solomon Islands, Sri Lanka, Tokelau Islands, Tonga, Tuvalu, Vanuatu, Western Samoa
Rest of the World	

Major changes from the listing between 1995 and 2005 by year are:

From 1995 - Cameroon and Mozambique moved from the *Rest of the World* to *New Commonwealth - Rest of Africa*; Austria, Finland and Sweden moved from *Rest of Europe* to *Other European Union*.

From 1997 - Hong Kong moved from *New Commonwealth - Far East* to the *Rest of the World*; Fiji moved from the *Rest of the World* to the *Rest of the New Commonwealth*.

From 2003 - Zimbabwe moved from *New Commonwealth - Rest of Africa* to the *Rest of the World*.

From 2004 - Cyprus and Malta moved from *New Commonwealth - Mediterranean* to *Other European Union*; Gibraltar moved from *New Commonwealth - Mediterranean* to the *Rest of the New Commonwealth*; Czech Republic, Estonia, Hungary, Latvia, Lithuania, Poland, Slovakia and Slovenia moved from *Rest of Europe* to *Other European Union*.

From 2005 - No major changes in the country listings occurred in 2005.

2.8   Stillbirths

In Section 41 of the Births and Deaths Registration Act 1953, a stillbirth is defined as 'a child which has issued forth from its mother after the twenty-eighth week of pregnancy and which did not at any time after being completely expelled from its mother breathe or show other signs of life'. This definition was used up to 30 September 1992.

On 1 October 1992 the Stillbirth (Definition) Act 1992 came into force, altering the above definition of a stillbirth to 24 or more weeks completed gestation. Figures for stillbirths from 1993 are thus not fully comparable with those for previous years. The effect of this change on figures for 1992 is analysed in the annual volume for that year.[8]

2.9   True birth order

When a birth is within marriage, information is obtained on the number of the mother's previous children, both live births and stillbirths. This allows determination of the *registration birth order* - that is, the number of previous live births plus the birth which has just occurred, counting only those births fathered by a previous or current husband. However, this measure is deficient for fertility statistics in two respects:

a.   at registration, the question on previous live births and stillbirths is not asked where the birth occurred outside marriage, and

b. at the registration of births and stillbirths occurring within marriage, previous live births occurring outside marriage and where the woman had never been married to the father are not counted. However, because of the ambiguous nature of the question (see **Annex A**) it is possible that births outside marriage where the woman subsequently married the father may not always be included.

The proportion of births occurring outside marriage has risen steadily in recent years. To allow for this, the information collected on birth order has been supplemented to give estimates of overall or *true birth order* - that is, a measure which includes births both within and outside marriage. These estimates are obtained from details provided by the General Household Survey (GHS).

The following example of a hypothetical birth history helps to illustrate the relationship between true birth order, marital birth order and birth order collected at registration:

Birth history	True birth order	Registration birth order	Marital birth order
First birth while cohabiting with man A	1	Not recorded	Not applicable
Second birth while married to man B	2	1	1
Third birth while cohabiting with man C	3	Not recorded	Not applicable
Fourth birth after marriage to man C	4	3	2

Previous volumes of *Birth statistics* used information from the GHS for 1986–1996, 1998 and 2000 to produce the estimates of true birth order. For the 2004 and 2005 volumes **Tables 1.7**, **10.3** and **10.5** use information from the GHS for the years 2001-2004. The method of estimation is described in a *Population Trends* article.[9]

Table 1.7 shows mean ages for each true birth order calculated in two ways:

a. Unstandardised means use only numbers of true order births and single year of age of mother.

b. Standardised means use rates per 1,000 female population by true birth order and single year of age of mother. This serves to eliminate the effect of year to year changes in the age-structure of the female population.

2.10 Births within marriage, and sole and joint registration

A birth within marriage is that of a child born to parents who were lawfully married to each other either:

a. at the date of the child's birth, or

b. when the child was conceived, even if they later divorced or the father died before the child's birth.

Only for a birth within marriage will the registrar enter on the draft entry in the register (*Form 309* - **Annex A** or *Form 308* - **Annex B**) confidential particulars relating to the date of the parents' marriage, whether the mother has been married more than once, and the number of the mother's previous live born and stillborn children - see section 1.2.

Births occurring outside marriage may be registered either jointly or solely. A joint registration records details of both parents, and requires them both to be present. A sole registration records only the mother's details. In a few cases a joint registration is made in the absence of the father if an affiliation order or statutory declaration is provided.

Information from the birth registrations is used to determine whether the mother and father jointly registering a birth outside marriage were usually resident at the same address at the time of registration - see **Table 3.10.** Births with both parents at the same address are identified by a single entry for the informant's usual address, while different addresses are identified by two entries.

2.11 Rates

In this volume, fertility rates have been calculated using the most up-to-date consistent mid-year estimates of the female population, based on the 2001 Census. See section 2.1 and **Appendix Table 1**.

The most commonly used rates are described below[10]:

Crude birth rate

This is the simplest overall measure of fertility in the population, given by the number of live births in a year per 1,000 mid-year population. It is unsophisticated since it takes no account of the composition of the population, in particular the age and sex distribution.

It is given by $(B/P) \times 1,000$
where B = total live births in the year, and
P = mid-year population.

In this volume it is used in **Tables 1.1b** and **1.3**.

General fertility rate (GFR)

This is an easily calculated measure of current fertility levels, and denotes the number of live births per 1,000 women aged 15-44. However, it makes no allowance for different sized cohorts of women at childbearing ages.

It is given by $(B/P^f_{15-44}) \times 1,000$
where B = total live births in the year, and
P^f_{15-44} = female population aged 15-44.

In this volume it is used in **Tables 1.1b**, **7.1** and **7.2.**

Age-specific fertility rates (ASFRs)

ASFRs are a measure of fertility specific to the age of the mother, and are useful for comparing the reproductive behaviour of women at different ages. They are calculated by dividing the number of live births to mothers of each age group by the number of females in the population of that age and then expressed per 1,000 women in the age group. They can be calculated for single ages, but are usually calculated for five-year age groups in the reproductive age range, from under 20 up to 45 and over. They provide the basis for a detailed analysis of fertility levels by age of mother when giving birth.

The ASFR based on five-year age groups is given by
$$F_a = (B_a/P^f_a) \times 1,000$$
where F_a = age-specific birth rate for age-group a,
B_a = live births to women in age-group a,
P^f_a = female population in age-group a, and
a = age-group under 20, ... , 45 and over.

For the groups under 20 and 45 and over, the female populations used are women aged 15-19, and women aged 45-49 respectively.

ASFRs are used in **Tables 7.3** and **7.4** of this volume.

Total fertility rate (TFR)

In this volume, for most tables the TFR is derived by summing single-year age-specific fertility rates over all ages within the childbearing lifespan. It is a measure independent of variations in the age distribution of women of childbearing age. It may be interpreted as representing the completed fertility of a synthetic cohort of women - that is, the average number of live children that a woman would bear if the female population experienced the age-specific fertility rates of the calendar year in question throughout their childbearing lifespan.

From the above the $\text{TFR} = \sum\limits_{a=\text{under }16}^{a=44 \text{ and over}} F_a$

where $F_a = B_a/P^f_a$
and B_a = live births to women in age-group a,
P^f_a = female population in age-group a, and
a = ages under 16, 16, 17, ... , 42, 43, 44 and over.

For the groups under 16 and 44 and over, the female populations used are women aged 15, and women aged 44 respectively.

The TFR is used in **Tables 1.4, 1.9, 2.1, 2.3, 7.1, 7.2** and **9.5.**

Gross reproduction rate (GRR)

The GRR is the sum of age-specific fertility rates for female births only, calculated in the same way as the TFR. It represents the average number of live daughters that a woman would bear in her life if the female population experienced current age-specific fertility rates based on female births throughout their childbearing period.

The GRR is used in **Table 1.4.**

Net reproduction rate (NRR)

The NRR is similar to the GRR, but allows for the effect of mortality to women of childbearing age. Not all women survive to the end of the possible reproductive period. It represents the average number of live daughters that a woman would bear if the female population experienced current age-specific fertility rates and female survival rates throughout their childbearing period.

The NRR is used in **Table 1.4.**

Average family size

Average family size is presented in this volume for women by year of birth and age, in tables on cohort fertility. For each cohort (i.e. women born in a particular year) it represents the number of births per woman, and is shown by birth order for births within marriage. Thus in **Table 10.4** the cohort born in 1951 had, between ages 20 and 24, borne on average 0.65 children per woman. Further, 0.59 were within marriage, 0.31 were first births per woman, 0.21 second births, and so on.

Stillbirth rate

The stillbirth rate is defined as the number of stillbirths per 1,000 live births and stillbirths, and is used in **Tables 1.2, 7.6** and **8.3.**

Sex ratio

Expressed as males per 1,000 females, most often for live births, but also for stillbirths. It is used here in **Tables 1.1b, 1.2** and **4.3.**

Other rates used in this volume include:

Live births within marriage per 1,000 married women, by age - **Tables 1.1b** and **3.1b.**

Live births within marriage per 1,000 married men, by age - **Tables 3.3** and **3.5.**

Live births outside marriage per 1,000 single, widowed and divorced women, by age - **Tables 1.1b** and **3.1b.**

Live births outside marriage per 1,000 live births - **Tables 1.1b, 7.1** and **7.2.**

Paternities within marriage per 1,000 married men, by age - **Table 3.5.**

Stillbirths within marriage per 1,000 married men, by age - **Table 3.5.**

Maternities with multiple births per 1,000 total maternities, by age - **Table 6.1b.**

Maternities within marriage with multiple births, per 1,000 maternities within marriage, by age - **Tables 6.1b** and **6.3.**

Maternities outside marriage with multiple births, per 1,000 maternities outside marriage, by age - **Table 6.1b.**

2.12 Accuracy of information

The accuracy of information contained in the draft birth entry (*Form 309* for a live birth or *Form 308* for a stillbirth - **Annex A** and **B** respectively) is the responsibility of the informant(s) - usually the mother, or the mother and father where the registration is a joint one outside marriage. Wilfully supplying false information may render the informant(s) liable to prosecution for perjury.

It is believed that in general the information supplied by the informant(s) is correct. Computerised internal consistency checks are applied to each record to eliminate, as far as possible, errors made in the supply and recording of information on births.

There are a few, very small, known errors in the database each year which it is not possible to correct. Their effects on the statistics are explained in relevant sections of these notes - for instance, duration of marriage in section 3.4.

2.13 Historical information

The formal registration of live births commenced on 1 July 1837, while stillbirths have been registered only since 1 July 1927. Confidential particulars for statistical purposes have been ascertained since 1 July 1938, under the Population Statistics Act of that year. From the later date, it has also been possible to routinely distinguish multiple births.

The Population Statistics Act 1960, effective from 1 January 1961, added a question on father's date of birth to the confidential particulars requested in the case of births within marriage. This applied also to births outside marriage where the father's name is entered in the register. Questions on father's and mother's place of birth were introduced on 1 April 1969 by the Registration of Births, Deaths and Marriages Regulations 1968.

As noted in section 2.8, the Stillbirth (Definition) Act of 1992 altered the definition of the gestation period for a stillbirth from 28 to 24 completed weeks.

2.14 Legislation

The main statutes concerning birth registration and provision of information on births are given below:

- **Census Act 1920,** which in Section 5 provides for the collection and publication of statistical information on the population by the Registrar General.

- **Population Statistics Act 1938,** which deals with the statistical information collected at registration.

- **Births and Deaths Registration Act 1953,** which covers all aspects of the registration of births and deaths.

- **Registration Service Act 1953,** which in Section 19 requires the Registrar General to provide annual abstracts of live births and stillbirths.

- **Population Statistics Act 1960,** which makes further provision for collecting statistical detail at registration.

- **Abortions Act 1967,** which permits termination of pregnancy by a registered practitioner, subject to certain conditions.

- **Registration of Births, Deaths and Marriages Regulations 1968,** which added questions on father's and mother's place of birth to the details requested at registration.

- **National Health Service Act 1977,** which requires notification of a birth to the health authority where the birth occurred.

- **Stillbirth (Definition) Act 1992,** which altered the definition of a stillbirth to 24 or more weeks completed gestation, instead of the previous definition of 28 or more weeks completed gestation.

- **Health Act 1999,** a section of which includes specific provision for the supply of information on individual births to the National Health Service.

2.15 Symbols and conventions

In this volume:

:	denotes not appropriate
0	denotes less than 0.5
-	denotes nil
*	denotes not available (to protect confidentiality)

Where data are not yet available, cells in tables are left blank. Rates and percentages calculated from fewer than 20 events (for example **Table 3.5**) are distinguished by italic type as a warning to users that their reliability as a measure may be affected by the small number of events.

Figures in some tables in this publication may not add precisely due to rounding or suppression.

Certain tables (for example **Tables 3.7, 4.2, 4.3** and **6.4**) showing a combination of information on stillbirths, and data items collected under the Population Statistics Act have been aggregated to protect confidentiality.[11]

2.16 Further information

Requests for births data, as well as background information on this volume, on unpublished VS tables, and on data quality, should be made to:

Vital Statistics Outputs Branch
Office for National Statistics
Segensworth Road
Titchfield
Fareham
Hants PO15 5RR
Telephone: 01329 813758; Fax: 01329 813548
email: vsob@ons.gsi.gov.uk

3   Notes on tables

3.1   Seasonality *(Tables 2.1 to 2.5)*

Seasonally adjusted numbers of live births are obtained using the X-11 ARIMA package developed by Statistics Canada. This software adjusts the monthly totals to eliminate seasonal fluctuations present in the latest 12 years of data, controlling them to the observed annual totals. The annual TFRs shown in **Tables 2.1 and 2.3** have been calculated using monthly and quarterly totals (adjusted for the number of days in those periods). The addition of a new year's data and the dropping of the oldest year means that Tables 2.1 and 2.3 will show slightly different figures in this volume over the whole 10 year series figures for those years in previous volumes. The population denominators are calculated for each month and quarter by apportioning the difference between one mid-year estimate and the next according to the time interval from the first. So the mid-November 2000 estimate is (mid-2001) − (mid-2000) x 4.5/12 + (mid-2000) because mid-November is 4.5 months away from end-June. This method has been used since the 2001 volume.

3.2   Age of parents *(Tables 3.1 to 3.10)*

The mother's or father's date of birth is recorded and translated into the age at the birthday **preceding** the date of the child's birth. This age is often termed age last birthday. Special checks are carried out on those dates of birth which imply that the age of the mother is over 50 years. While most dates of birth are confirmed, these extreme values tend to occur more often among women born outside England and Wales - see **Tables 9.1** to **9.6** - and may result from age misreporting.

If either the mother's date of birth or the father's date of birth (when applicable) is not given, an age is imputed from a similar record with completely stated and otherwise matching particulars. In 2005, the mother's date of birth was not stated for 1.1 per cent of all live births, and the father's date of birth was not stated for 1.2 per cent of all live births where father's details were present.

3.3   Previous live-born children *(Tables 4.1 to 4.3)*

Information on previous live-born children is available only for women having a birth within marriage. It denotes the number of previous live-born children by the present and any former husband, as stated at registration. This information is also used to determine true birth order - see section 2.9.

If the number of previous live-born children is not given, a value is imputed from a similar record with completely stated and otherwise matching particulars. In 2005, 0.1 per cent of all live births within marriage had this variable imputed. Of all stillbirths within marriage, 15.1 per cent had this variable imputed.

3.4   Duration of marriage *(Tables 5.1 to 5.3)*

Pre-maritally conceived live births are, by convention, taken to be those where the calculated duration of marriage is up to and including seven completed months. At registration only the month and year of marriage are recorded, so the calculation relates to the interval in completed months between the middle of the month of marriage and the date of the child's birth. Other durations of marriage are calculated in a similar way.

If the date of marriage is not given, a value for the date of marriage is imputed from a similar record with completely stated and otherwise matching particulars. In 2005, the year of marriage was not stated for 1.4 per cent of all live births within marriage. For women who have been married more than once, duration refers to the length of the current marriage.

In **Table 1.8**, the percentages shown are based on live births where the mother married in the year shown, and the birth was within eight months of marriage, and on first marriages in the same year. Thus the percentages (by age) for year of marriage 2004 would be based on (a) the number of live births in 2004 or 2005 where the mother conceived in 2004 and gave birth within eight months of marriage, and (b) the number of first marriages occurring in 2004.

3.5   Multiple births *(Tables 6.1 to 6.4)*

Multiple births arising from a single pregnancy are counted as one maternity or paternity, although each child born is reckoned separately in analyses of birth statistics. In tables analysing births by the number of previous live-born children, multiple births are counted as if they had occurred separately - in the case of twins for instance, as one first and one second birth.

3.6 Birthweight *(Tables 7.5 and 7.7)*

Birthweight is measured in grams, and is notified to the local health authority by the hospital where the birth took place, or by the midwife or doctor in attendance at the birth. These details are then supplied to the registrar. For stillbirths, details of the weight of the foetus are supplied on a certificate or notification by a doctor or midwife. The certificate or notification is then taken by an informant to the registrar.

In cases where no birthweight is recorded, the birth is included in the total 'all weights' but not distributed amongst the individual categories. (These categories may thus not add to the total in **Tables 7.5** and **7.7.**). In 2005, birthweight was not stated for 0.4 per cent of all live births, and for 1.6 per cent of stillbirths.

3.7 Place of confinement and area of occurrence *(Tables 8.1 to 8.3)*

Place of confinement is categorised as follows:

NHS establishments - generally hospitals, maternity units and maternity wings;

Non-NHS establishments - including private maternity units, military hospitals, and private hospitals;

At home - denoting the usual place of residence of the mother;

Elsewhere - including all locations not covered above: most of these are at a private residence not that of the mother, or are on the way to a hospital.

A birth is usually assigned to an area according to the usual residence of the mother at the time of birth, as stated at registration. However, a birth may take place in an area other than that of the mother's usual residence. **Table 8.2** shows whether a confinement takes place in the same area as the mother's usual area of residence, or if it occurred in a different area. Births which take place at home or elsewhere are not allocated a health area of occurrence. **Table 8.3** shows the numbers of maternities, live births and stillbirths occurring in NHS and non-NHS establishments, by area of occurrence.

3.8 Country of birth of mother *(Tables 9.1 to 9.6)*

For children born in England and Wales, the country of birth of parents has been recorded at birth registration since April 1969. However, it should be noted that birthplace does not necessarily equate with ethnic group. A fuller discussion of this subject can be found elsewhere.[12]

Country of birth groupings represent the Commonwealth and European Union as constituted in 2005 - see section 2.7.

3.9 Birth cohorts *(Tables 10.1 to 10.5)*

Birth statistics analysed by year of occurrence and by age of mother have been available since 1938. **Tables 10.1** to **10.5** show these statistics in cohort form - by the year of birth of the mother rather than the year of birth of the child. The years of birth shown are by necessity approximate since, prior to 1963, data are available only by calendar year of occurrence and age of mother at childbirth. For instance, women aged 32 giving birth to children in 2003 could have been born in either 1970 or 1971; for convenience, however, such women are here regarded as belonging to the 1971 cohort.

Tables 10.1 to **10.5** all refer to age in completed years. **Table 10.2** gives for a particular cohort (women born in a given year) the average number of live-born children after n completed years of age. When data become available for a given cohort from 15 to 45 completed years, then the figure shown after 45 completed years of age measures the average completed family size for women born in that cohort. This is calculated by summing the age-specific birth rates for that cohort shown in **Table 10.1** up to and including age n. For example, **Table 10.2** shows that women born in 1970 had given birth to 0.22 children on average after 20 completed years of age. This was calculated by adding the age-specific birth rates for the 1970 cohort in **Table 10.1** up to and including age 20 – that is, $(3+12+28+48+61+72)/1,000 = 0.22$.

3.10 Socio-economic classification as defined by occupation *(Tables 11.1 to 11.5)*

The information on occupation of the father is coded for only a sample of one in ten births. Combining this with the employment status, a code for socio-economic classification (or social class in previous volumes) may be derived. From 1991 to 2000 the occupation of the father was coded using the Standard Occupational Classification SOC90[13], and occupation codes were allocated to the Registrar General's Social Class.

The Standard Occupational Classification is revised every ten years and in 2001 SOC2000[14, 15] replaced SOC90. The coding of employment status also changed in 2001 to be consistent with the 2001 Census and SOC2000. Since 2001, the National Statistics Socio-economic Classification (NS-SEC)[16] has categorised the socio-economic classification of people, and has replaced the Registrar General's Social Class and the Socio-economic Group (SEG). SOC2000 and employment status are used to derive NS-SEC for births. The new classification is based not on skills but on employment conditions, which are now considered to be central to describing the socio-economic structure of modern societies.

NS-SEC has eight analytic classes, the first of which can be subdivided:

1. Higher managerial and professional occupations
   1.1 Large employers and higher managerial occupations
   1.2 Higher professional occupations
2. Lower managerial and professional occupations
3. Intermediate occupations
4. Small employers and own-account workers
5. Lower supervisory and technical occupations
6. Semi-routine occupations
7. Routine occupations
8. Never worked and long-term unemployed

Students, occupations not stated or inadequately described, and occupations not classifiable for other reasons are added as 'Not Classified'.

The sample figures in **Tables 11.1** to **11.5** have been grossed-up to agree with known totals derived from the 100 per cent processing of birth registrations by mother's age and previous live-born children in **Table 4.1a**. This ensures consistency with sub-totals, and improves the quality of sample estimates. **Appendix Tables 3** and **4** show standard errors for selected numbers of births and percentages. Thus, if the estimated grossed-up number in a particular category was 50,000, then from **Appendix Table 3** the standard error of that estimate would be approximately 640. Based on statistical theory, this means that for the type of distribution being considered there is about a 95 per cent chance that the 'true' number in the population lies within two standard errors of the estimates. This true number is that which would have been obtained had all the information been collected, rather than a one in ten sample.

In this example, the 95 per cent confidence interval would be:

50,000 ± 1,300, or 48,700 to 51,300.

In other words, we could say that we are 95 per cent confident that the true value, if we had collected all the information instead of a 10 per cent sample, lies somewhere between 48,700 and 51,300.

3.11 Birth intervals *(Table 11.3)*

Figures in **Table 11.3** showing median birth intervals are produced using two separate sources. The median interval between marriage and first birth is derived from births registration data. It relates to births occurring in England and Wales within marriage only and to the first child that

is fathered by the present or any former husband. For remarried women, the interval is measured from the date of the current marriage. Where the first maternity is a multiple birth, the interval is measured from marriage for each resulting child.

The other part of the table shows the median intervals between first, second, third and fourth births. It is derived by the Inland Revenue from a 5 per cent sample of new claims for child benefit from all births occurring in the United Kingdom - whether within or outside marriage. A zero interval is assumed for births resulting from a multiple maternity.

References

1. OPCS (1987). *Birth statistics 1837-1983*, series FM1 no 13.
2. ONS (2006). *Birth statistics 2004*, series FM1 no.33 (revised).
3. Statistics Canada (2005). *CANCEIS User's Guide: Canadian Census Edit and Imputation System*, CANCEIS Development Team, Social Survey Methods Division.
4. ONS (1999). *Birth statistics 1998*, series FM1 no 27.
5. ONS (2000). *Birth statistics 1999*, series FM1, no 28.
6. ONS (2006). *Key population and vital statistics - local and health authority areas 2004*, series VS no 31, PP1 no 27.
7. ONS (2001) *Birth statistics 2000*, series FM1 no 29.
8. OPCS (1994). *Birth statistics 1992*, series FM1 no 21.
9. Smallwood, S. (2002). 'New estimates of trends in birth order in England and Wales'. *Population Trends* 108, pp 32-48.
10. Shryock, HS. and Siegel, JS. (1973). *The methods and materials of demography*, chapter 16. (US Government Printing Office, Washington DC, 1973).
11. ONS (2005) *ONS Policy on protecting confidentiality within birth and death statistics*. See website: www.statistics.gov.uk/statbase/Product.asp?vlnk=5678
12. Shaw, C. (1988). 'Components of growth in the ethnic minority population'. *Population Trends* 52, pp 26-30.
13. OPCS (1990). *Standard Occupational Classification: volumes 1-3*, HMSO: London.
14. ONS (2000). *Standard Occupational Classification 2000: Volume 1. Structure and descriptions of unit groups*, TSO: London.
15. ONS (2000). *Standard Occupational Classification 2000: Volume 2. The coding index*, TSO: London.
16. Rose, D and O'Reilly, K (1998). *The ESRC Review of Government Social Classifications*, ESRC & ONS: Swindon.

Glossary

Abortion

The legal termination of a pregnancy under the 1967 Abortion Act.

Age-Specific Fertility Rate (ASFR)

The number of live births to mothers of a particular age per 1,000 women in that age group. Useful for comparing fertility of women at different ages or women with the same ages in different populations.

Annual Reference Volume (ARV)

ARVs are yearly publications produced by ONS for a variety of topics (for example, FM1 for Birth statistics).

Cohort

A specific group of people, in this case, those born during a particular year. Analysis using cohorts considers the experience of that group of people over time.

Conception

ONS uses the definition - a pregnancy of a woman which leads either to a maternity or an abortion.

Crude Birth Rate

The number of live births in a year per 1,000 mid-year population.

General Fertility Rate (GFR)

The number of live births per 1,000 women aged 15-44. Measure of current fertility levels.

General Household Survey (GHS)

The GHS is a continuous survey carried out by ONS, collecting information on a range of topics from people living in private households in Great Britain.

General Register Office (GRO)

The GRO is responsible for ensuring the registration of all births, marriages and deaths that have occurred in England and Wales since 1837 and for maintaining a central archive.

Gross Reproduction Rate (GRR)

The sum of age-specific fertility rates for female births only. The average number of live daughters that a woman would bear in her life, if the female population experienced current ASFRs based on female births throughout their childbearing years.

Health Statistics Quarterly

A quarterly ONS publication that covers mortality and health information, including articles and reports on conceptions.

Informant

The person(s), normally one or both parents, who provide the registrar with the information required at the registration of a birth.

Imputation

A method used to add information to an incomplete birth record, using the details from another similar but complete record.

Joint Registration

A birth outside marriage registered by both the mother and father of the child. Both parents' details are recorded and both must be present at the registration.

Live Birth

A baby showing signs of life at birth.

Maternity

A confinement resulting in the birth of one or more live-born or stillborn children. Therefore, the number of maternities (and **paternities**) is less than the total number of live births and stillbirths.

Mean	A common measure of the average. The values are summed and then divided by the total number of observations.
Median	Statistical term for the value for which half the data are above and half are below. An alternative measure to the mean.
Multiple Births	A single maternity resulting in two or more births.
National Statistics Code of Practice	The principles and protocols followed and upheld by all those involved in producing National Statistics.
Natural Change	The change to the size of the population due to births and deaths (not taking into account the contribution of migration).
Net Reproduction Rate (NRR)	Similar to the GRR, but also takes into account the effect of mortality to women of childbearing age.
News Release	Once an annual dataset from a particular source such as birth registration has been quality assured, its first publication is a press, or news, release, which details the main findings.
NS-SEC	National Statistics Socio-economic Classification categorises the socio-economic classification of people, and has replaced the Registrar General's Social Class and the Socio-economic Group (SEG).
Occurrences	Births which occur in a given period, for example a calendar year.
ONS	Office for National Statistics.
OPCS	Office of Population Censuses and Surveys - joined with Central Statistical Office in 1996 to become ONS.
Parity	The number of live births a woman has had. A woman who has one child has a parity of one. See Registration Birth Order and True Birth Order.
Place of Confinement	Place where a birth occurs.
Population Statistics Act (PSA)	This Act makes provision for certain information to be collected at the registration of the birth for statistical use. This information is confidential and is not entered on the register.
Population Trends	A quarterly ONS publication that covers population and demographic information, including articles and reports on births.
Ratio	A measure of the relative size of two variables expressed as a proportion.
Registrar	Statutory officer responsible for the registration of births, deaths and marriages.
Registrar General	Statutory appointment with responsibility for the administration of the registration Acts in England and Wales, and other related functions as specified by the relevant legislation.
Registration Birth Order	The number assigned to a birth based on the number of previous live births to that mother, counting only those births fathered by her current or any previous husband(s).

Registration Officer	Generic term for registrar, superintendent registrar and additional registrars.
Registrations	Births that were registered in a particular accounting period, even though some may have occurred in an earlier accounting period.
Reports	Regular articles in *Population Trends* and *Health Statistics Quarterly* on births and conceptions.
Seasonality	The effect of seasonal fluctuations to monthly and quarterly births figures. Monthly totals are adjusted in **Tables 2.1** and **2.3** to eliminate this effect.
Single Men/Women	Persons who have never been married.
Singleton	A live birth or stillbirth which is the sole live birth or stillbirth born of a maternity.
SOC2000	Standard Occupational Classification 2000 is the current occupational classification. SOC2000 codes, details of employment status and size of organisation are required for the derivation of NS-SEC. See NS-SEC.
Sole Registration	A birth outside of marriage registered only by the mother. No information on the father is recorded.
Standard Error	A measure of the sampling variation occurring by chance when only part of the total population has been selected, for example, father's occupation.
Standardised Mean Age	The average age (for example, at birth or marriage) of the population in question calculated to take into account the changing distribution of that population by age and other factors over time. This mean should be used when analysing trends.
Stillbirth	A child that has issued forth from its mother after the 24th week of pregnancy, and that did not at any time after being completely expelled from its mother breathe or show any signs of life.
Superintendent Registrar	Statutory officer with responsibilities relating to marriage and other registration functions, as specified in the relevant legislation.
Total Fertility Rate (TFR)	The sum of the age-specific fertility rates. The average number of live children that a woman would bear if the female population experienced the ASFRs of the calendar year in question throughout their childbearing lifespan.
True Birth Order	The number assigned to a birth based on the number of previous live births to that mother, counting all births inside or outside of marriage.
Unstandardised Mean Age	The average age (for example, at birth or marriage) of the population in question, calculated as the actual average for a particular year. It does not take into account the changing distribution of that population over time. This measure should be used when requiring a mean for particular year.
VSOB	Vital Statistics Output Branch (at ONS).

Table 1.1 Live births: occurrence within/outside marriage and sex, 1995-2005   **England and Wales**
a. numbers

Year	All			Within marriage			Outside marriage		
	Total	Males	Females	**Total**	Males	Females	**Total**	Males	Females
1995	**648,138**	332,188	315,950	**428,189**	219,475	208,714	**219,949**	112,713	107,236
1996	**649,485**	333,490	315,995	**416,822**	214,542	202,280	**232,663**	118,948	113,715
1997	**643,095**	329,577	313,518	**404,873**	207,199	197,674	**238,222**	122,378	115,844
1998	**635,901**	325,903	309,998	**395,290**	202,762	192,528	**240,611**	123,141	117,470
1999	**621,872**	319,255	302,617	**379,983**	194,935	185,048	**241,889**	124,320	117,569
2000	**604,441**	309,625	294,816	**365,836**	187,367	178,469	**238,605**	122,258	116,347
2001	**594,634**	304,635	289,999	**356,548**	182,899	173,649	**238,086**	121,736	116,350
2002	**596,122**	306,063	290,059	**354,090**	181,452	172,638	**242,032**	124,611	117,421
2003	**621,469**	318,428	303,041	**364,244**	186,730	177,514	**257,225**	131,698	125,527
2004	**639,721**	328,340	311,381	**369,997**	189,861	180,136	**269,724**	138,479	131,245
2005	**645,835**	330,600	315,235	**369,330**	189,388	179,942	**276,505**	141,212	135,293

Table 1.1 Live births: occurrence within/outside marriage and sex, 1995-2005   **England and Wales**
b. rates and sex ratios

Year	Crude birth rate: all births per 1,000 population of all ages[1]	General fertility rate: all births per 1,000 women aged 15-44[1]	Births within marriage per 1,000 married women aged 15-44[1]	Births outside marriage per 1,000 single, widowed and divorced women aged 15-44[1]	Births outside marriage per 1,000 total births	Sex ratio: male births per 1,000 female births		
						All	Within marriage	Outside marriage
1995	12.6	60.5	82.7	39.7	339.4	1,051	1,052	1,051
1996	12.6	60.6	82.2	41.2	358.2	1,055	1,061	1,046
1997	12.5	60.0	81.6	41.3	370.4	1,051	1,048	1,056
1998	12.3	59.2	81.3	40.9	378.4	1,051	1,053	1,048
1999	12.0	57.8	79.6	40.4	389.0	1,055	1,053	1,057
2000	11.6	55.9	77.9	39.0	394.8	1,050	1,050	1,051
2001	11.4	54.7	77.3	38.0	400.4	1,050	1,053	1,046
2002	11.3	54.7	78.8	37.7	406.0	1,055	1,051	1,061
2003	11.8	56.8	83.3	39.1	413.9	1,051	1,052	1,049
2004	12.1	58.2	86.8	40.1	421.6	1,054	1,054	1,055
2005	12.1	58.4	88.8	40.1	428.1	1,049	1,052	1,044

1 See sections 2.1 and 2.11.

**Table 1.2 Stillbirths (numbers, rates and sex ratios):
occurrence within/outside marriage and sex, 1995-2005**   **England and Wales**

Year	Numbers									Rates			
	All			Within marriage			Outside marriage			Stillbirths per 1,000 live births and stillbirths[1]	Sex ratio: male births per 1,000 female births		
	Total	Males	Females	**Total**	Males	Females	**Total**	Males	Females		All	Within marriage	Outside marriage
1995	**3,600**	1,912	1,688	**2,224**	1,197	1,027	**1,376**	715	661	5.5	1,133	1,166	1,082
1996	**3,539**	1,808	1,731	**2,114**	1,057	1,057	**1,425**	751	674	5.4	1,044	1,000	1,114
1997	**3,439**	1,801	1,638	**2,010**	1,070	940	**1,429**	731	698	5.3	1,100	1,138	1,047
1998	**3,417**	1,822	1,595	**1,966**	1,054	912	**1,451**	768	683	5.3	1,142	1,156	1,124
1999	**3,305**	1,727	1,578	**1,875**	973	902	**1,430**	754	676	5.3	1,094	1,079	1,115
2000	**3,203**	1,731	1,472	**1,830**	979	851	**1,373**	752	621	5.3	1,176	1,150	1,211
2001	**3,159**	1,725	1,434	**1,771**	936	835	**1,388**	789	599	5.3	1,203	1,121	1,317
2002	**3,372**	1,807	1,565	**1,896**	1,030	866	**1,476**	777	699	5.6	1,155	1,189	1,112
2003	**3,612**	1,885	1,727	**1,977**	1,048	929	**1,635**	837	798	5.8	1,091	1,128	1,049
2004	**3,686**	1,950	1,736	**1,982**	1,039	943	**1,704**	911	793	5.7	1,123	1,102	1,149
2005	**3,483**	1,796	1,687	**1,861**	953	908	**1,622**	843	779	5.4	1,065	1,050	1,082

1 See section 2.11.

Table 1.3   Natural change in population   **England and Wales**
(numbers and rates[1]), 1995-2005

Year	Numbers			Rates per 1,000 population of all ages		
	Live births	Deaths	Natural change: live births minus deaths	Live births (crude birth rate)	Deaths (crude death rate)	Natural change
1995	648,138	569,683	78,455	12.6	11.1	1.5
1996	649,485	560,135	89,350	12.6	10.9	1.7
1997	643,095	555,281	87,814	12.5	10.8	1.7
1998	635,901	555,015	80,886	12.3	10.7	1.6
1999	621,872	556,118	65,754	12.0	10.7	1.3
2000	604,441	535,664	68,777	11.6	10.3	1.3
2001	594,634	530,373	64,261	11.4	10.1	1.2
2002	596,122	533,527	62,595	11.3	10.1	1.2
2003	621,469	538,254	83,215	11.8	10.2	1.6
2004	639,721	512,541	127,180	12.1	9.7	2.4
2005	645,835	512,541	133,143	12.1	9.6	2.5

1 See sections 2.1 and 2.11.

Table 1.4   Total fertility, gross and net   **England and Wales**
reproduction rates[1], 1995-2005

Year	Total fertility rate (TFR)	Gross reproduction rate (GRR)	Net reproduction rate (NRR)
1995	1.72	0.84	0.84
1996	1.74	0.85	0.84
1997	1.73	0.84	0.84
1998	1.72	0.84	0.84
1999	1.70	0.83	0.82
2000	1.65	0.81	0.80
2001	1.63	0.81	0.81
2002	1.65	0.82	0.81
2003	1.73	0.84	0.84
2004	1.78	0.87	0.86
2005	1.80	0.88	0.87

1 See sections 2.1 and 2.11.

Table 1.5   Marriages (numbers and rates[1]): sex, 1994-2004   **England and Wales**

Year	Number of marriages			Marriage rates				
	All	Single men	Single women	Crude marriage rate - all persons marrying per 1,000 population of all ages	Males marrying per 1,000 single, widowed and divorced males aged 16 and over	Females marrying per 1,000 single, widowed and divorced females aged 16 and over	Single men marrying per 1,000 single males aged 16 and over	Single women marrying per 1,000 single females aged 16 and over
1994	291,069	206,077	206,332	11.4	36.3	30.6	34.3	41.6
1995	283,012	198,208	198,603	11.0	34.7	29.3	32.4	39.3
1996	278,975	193,306	192,707	10.9	33.6	28.5	31.1	37.3
1997	272,536	188,268	188,457	10.6	32.3	27.5	29.7	35.6
1998	267,303	186,329	187,391	10.3	31.1	26.6	28.9	34.7
1999	263,515	184,266	185,328	10.1	30.1	25.8	28.0	33.5
2000	267,961	186,113	187,717	10.3	30.1	25.9	27.7	33.2
2001	249,227	175,721	177,506	9.5	27.4	23.7	25.5	30.6
2002	255,517	179,052	180,605	9.7	27.4	23.9	25.3	30.3
2003	270,109	189,470	191,170	10.2	28.2	24.8	26.1	31.2
2004	273,069	191,956	194,348	10.3	27.8	24.6	25.7	30.8

Note: These figures relate only to marriages solemnised in England and Wales.
1 See section 2.1.

Table 1.6 Mean age of all women at marriage and of mothers at live birth, 1995-2005   **England and Wales**

Year	Mean age at marriage		Mean age at live birth						
	All brides	Single women	All births	Births outside marriage	Births within marriage				
					All birth orders	First birth	Second birth	Third birth	Fourth birth
1995	28.1	26.0	**28.5**	26.0	29.8	28.5	30.0	31.1	32.0
1996	31.1	27.2	**28.6**	26.1	30.1	28.8	30.3	31.3	32.2
1997	31.4	27.5	**28.8**	26.2	30.3	29.0	30.5	31.5	32.4
1998	31.6	27.7	**28.9**	26.3	30.5	29.2	30.7	31.8	32.6
1999	31.8	28.0	**29.0**	26.4	30.6	29.3	30.9	32.0	32.7
2000	32.1	28.2	**29.1**	26.5	30.8	29.6	31.1	32.1	32.8
2001	32.2	28.4	**29.2**	26.7	30.9	29.6	31.2	32.2	33.0
2002	32.6	28.7	**29.3**	26.8	31.0	29.7	31.4	32.3	33.1
2003	32.9	28.9	**29.4**	26.9	31.2	29.9	31.5	32.5	33.1
2004	33.1	29.1	**29.4**	27.0	31.2	30.0	31.6	32.5	33.2
2005			**29.5**	27.0	31.3	30.1	31.7	32.5	33.3

Note: The mean ages shown in this table are unstandardised and therefore take no account of the structure of the population by age, marital status or parity.

Table 1.7a Mean age of mother by birth order, 1995-2005   **England and Wales**

Year	All births	True birth order[1]			
		First	Second	Third	Fourth
1995	28.5	26.6	29.0	30.3	31.2
1996	28.6	26.7	29.1	30.5	31.3
1997	28.8	26.8	29.3	30.8	31.5
1998	28.9	26.9	29.4	31.0	31.7
1999	29.0	27.0	29.5	31.0	31.8
2000	29.1	27.2	29.7	31.1	31.8
2001	29.2	27.2	29.8	31.2	31.9
2002	29.3	27.4	29.9	31.2	31.9
2003	29.4	27.5	29.9	31.3	32.0
2004	29.4	27.5	30.0	31.3	32.0
2005	29.5	27.6	30.0	31.4	32.1

Note: The mean ages shown in this table are unstandardised and therefore take no account of the structure of the population by age, marital status and parity.
1 See section 2.9.

Table 1.7b Mean age of mother standardised for age[1], 1995-2005   **England and Wales**

Year	All births	True birth order[2]			
		First	Second	Third	Fourth
1995	28.2	26.1	28.7	30.3	31.3
1996	28.2	26.1	28.8	30.4	31.3
1997	28.3	26.2	28.8	30.6	31.4
1998	28.3	26.3	28.9	30.7	31.5
1999	28.4	26.4	29.0	30.6	31.5
2000	28.5	26.5	29.1	30.6	31.4
2001	28.6	26.6	29.2	30.7	31.5
2002	28.7	26.8	29.3	30.7	31.5
2003	28.8	27.0	29.4	30.8	31.5
2004	28.9	27.1	29.5	30.8	31.6
2005	29.0	27.3	29.6	30.9	31.6

1 Standardised for the age distribution of the population. This measure is more appropriate for use when analysing trends or making comparisons between different geographies.
2 See section 2.9.

Table 1.8 Percentage of first marriages with a birth within   **England and Wales**
8 months of marriage: age of mother at marriage, 1994-2004[1]

Year of marriage	Woman's age at marriage				
	Under 20	20-24	25-29	30-44	Under 45
1994	22.5	9.9	8.3	11.1	10.2
1995	20.6	10.3	8.3	10.8	10.1
1996	23.1	11.0	8.5	10.5	10.5
1997	21.9	11.1	8.6	10.8	10.5
1998	22.7	11.4	8.3	10.4	10.4
1999	22.7	11.1	8.0	10.0	10.0
2000	19.9	10.9	7.5	8.8	9.3
2001	22.0	10.9	7.2	8.7	9.2
2002	22.1	11.0	7.5	8.9	9.4
2003	20.1	11.2	7.8	9.0	9.5
2004	20.5	11.4	7.8	8.5	9.4

1 See section 3.4.

Table 1.9 Total fertility rates[1]: occurrence within/outside   **England and Wales**
marriage, birth order[2] and age of mother, 1995-2005

Year	All live births	Live births outside marriage	Live births within marriage						Year	All live births	Live births outside marriage	Live births within marriage					
			All	Birth order								All	Birth order				
				First	Second	Third	Fourth	Fifth and later					First	Second	Third	Fourth	Fifth and later
All ages of mother at birth									**25-29**								
1995	**1.72**	0.62	1.10	0.43	0.41	0.17	0.06	0.04	1995	**0.54**	0.15	0.39	0.17	0.14	0.05	0.02	0.01
1996	**1.74**	0.66	1.08	0.42	0.40	0.17	0.06	0.03	1996	**0.53**	0.16	0.38	0.17	0.14	0.05	0.02	0.01
1997	**1.73**	0.68	1.05	0.41	0.39	0.17	0.06	0.03	1997	**0.52**	0.16	0.36	0.16	0.13	0.05	0.02	0.01
1998	**1.72**	0.69	1.03	0.41	0.38	0.16	0.05	0.03	1998	**0.51**	0.17	0.34	0.16	0.12	0.04	0.01	0.01
1999	**1.70**	0.70	1.00	0.40	0.37	0.15	0.05	0.03	1999	**0.49**	0.17	0.32	0.15	0.11	0.04	0.01	0.01
2000	**1.65**	0.69	0.96	0.39	0.35	0.15	0.05	0.03	2000	**0.47**	0.16	0.31	0.14	0.11	0.04	0.01	0.01
2001	**1.63**	0.69	0.95	0.38	0.35	0.14	0.05	0.03	2001	**0.46**	0.16	0.29	0.14	0.10	0.04	0.01	0.01
2002	**1.65**	0.70	0.95	0.39	0.35	0.14	0.05	0.03	2002	**0.46**	0.17	0.29	0.14	0.10	0.04	0.01	0.01
2003	**1.73**	0.74	0.99	0.41	0.36	0.14	0.05	0.03	2003	**0.48**	0.18	0.30	0.15	0.10	0.04	0.01	0.01
2004	**1.78**	0.77	1.01	0.43	0.37	0.14	0.05	0.03	2004	**0.49**	0.19	0.30	0.15	0.10	0.04	0.01	0.01
2005	**1.80**	0.79	1.02	0.43	0.36	0.14	0.05	0.03	2005	**0.50**	0.19	0.30	0.15	0.10	0.04	0.01	0.01
Under 20									**30-34**								
1995	**0.15**	0.13	0.02	0.02	0.00	0.00	0.00	0.00	1995	**0.44**	0.09	0.35	0.11	0.14	0.07	0.02	0.01
1996	**0.15**	0.14	0.02	0.02	0.00	0.00	0.00	0.00	1996	**0.45**	0.10	0.35	0.11	0.14	0.06	0.02	0.01
1997	**0.15**	0.14	0.02	0.01	0.00	0.00	0.00	0.00	1997	**0.45**	0.10	0.35	0.11	0.14	0.06	0.02	0.01
1998	**0.15**	0.14	0.02	0.01	0.00	0.00	0.00	0.00	1998	**0.45**	0.11	0.35	0.12	0.14	0.06	0.02	0.01
1999	**0.15**	0.14	0.02	0.01	0.00	0.00	0.00	0.00	1999	**0.45**	0.11	0.34	0.12	0.14	0.06	0.02	0.01
2000	**0.15**	0.13	0.02	0.01	0.00	0.00	0.00	0.00	2000	**0.44**	0.11	0.33	0.12	0.14	0.05	0.02	0.01
2001	**0.14**	0.13	0.02	0.01	0.00	0.00	0.00	0.00	2001	**0.44**	0.11	0.33	0.12	0.13	0.05	0.02	0.01
2002	**0.14**	0.12	0.02	0.01	0.00	0.00	0.00	0.00	2002	**0.45**	0.12	0.33	0.12	0.14	0.05	0.02	0.01
2003	**0.14**	0.12	0.01	0.01	0.00	0.00	0.00	0.00	2003	**0.48**	0.13	0.35	0.14	0.14	0.05	0.02	0.01
2004	**0.14**	0.12	0.01	0.01	0.00	0.00	0.00	0.00	2004	**0.50**	0.13	0.37	0.14	0.14	0.05	0.02	0.01
2005	**0.13**	0.12	0.01	0.01	0.00	0.00	0.00	0.00	2005	**0.51**	0.14	0.37	0.15	0.14	0.05	0.02	0.01
20-24									**35 and over**								
1995	**0.38**	0.21	0.17	0.09	0.06	0.02	0.00	0.00	1995	**0.21**	0.05	0.17	0.04	0.06	0.04	0.02	0.02
1996	**0.38**	0.22	0.16	0.09	0.06	0.02	0.00	0.00	1996	**0.22**	0.05	0.17	0.04	0.06	0.04	0.02	0.02
1997	**0.38**	0.22	0.15	0.08	0.05	0.02	0.00	0.00	1997	**0.23**	0.06	0.17	0.04	0.06	0.04	0.02	0.02
1998	**0.37**	0.22	0.15	0.08	0.05	0.02	0.00	0.00	1998	**0.24**	0.06	0.18	0.04	0.06	0.04	0.02	0.01
1999	**0.36**	0.22	0.14	0.08	0.05	0.01	0.00	0.00	1999	**0.24**	0.06	0.18	0.04	0.07	0.04	0.02	0.01
2000	**0.35**	0.22	0.13	0.07	0.05	0.01	0.00	0.00	2000	**0.25**	0.07	0.18	0.05	0.07	0.04	0.02	0.01
2001	**0.35**	0.22	0.13	0.07	0.04	0.01	0.00	0.00	2001	**0.25**	0.07	0.18	0.05	0.07	0.04	0.02	0.01
2002	**0.35**	0.22	0.13	0.07	0.04	0.01	0.00	0.00	2002	**0.26**	0.07	0.19	0.05	0.07	0.04	0.02	0.01
2003	**0.36**	0.23	0.12	0.07	0.04	0.01	0.00	0.00	2003	**0.28**	0.08	0.20	0.05	0.08	0.04	0.02	0.01
2004	**0.36**	0.24	0.12	0.07	0.04	0.01	0.00	0.00	2004	**0.30**	0.09	0.21	0.06	0.08	0.04	0.02	0.01
2005	**0.36**	0.24	0.12	0.06	0.04	0.01	0.00	0.00	2005	**0.31**	0.09	0.21	0.06	0.08	0.04	0.02	0.01

1 See sections 2.1 and 2.11.
2 See section 2.9.

Table 2.1   Live births (actual and seasonally adjusted numbers and rates[1]):   England and Wales
            quarter of occurrence, 1995-2005

Year	Total	Quarter ending				Total	Quarter ending			
		31 March	30 June	30 September	31 December		31 March	30 June	30 September	31 December
	Number (thousands)					**Total fertility rate (TFR)**				
	Live births					**Actual TFR**				
1995	**648.1**	158.5	164.7	167.4	157.5	**1.72**	1.70	1.75	1.76	1.66
1996	**649.5**	157.3	158.1	169.9	164.2	**1.74**	1.69	1.69	1.81	1.76
1997	**643.1**	158.1	163.3	164.9	156.8	**1.73**	1.72	1.76	1.76	1.68
1998	**635.9**	155.8	158.6	166.1	155.4	**1.72**	1.71	1.72	1.79	1.68
1999	**621.9**	152.1	157.3	160.1	152.4	**1.70**	1.68	1.71	1.73	1.65
2000	**604.4**	148.7	150.7	155.0	150.1	**1.65**	1.63	1.65	1.69	1.64
2001	**594.6**	145.5	148.8	153.0	147.4	**1.63**	1.62	1.64	1.67	1.61
2002	**596.1**	143.3	147.2	155.0	150.6	**1.65**	1.60	1.63	1.70	1.66
2003	**621.5**	147.4	155.2	162.9	156.0	**1.73**	1.66	1.73	1.80	1.72
2004	**639.7**	155.2	157.4	165.4	161.7	**1.78**	1.74	1.76	1.84	1.79
2005	**645.8**	154.3	159.8	170.2	161.7	**1.80**	1.77	1.78	1.88	1.79
	Seasonally adjusted live births[2]					**Seasonally adjusted TFR[2]**				
1995		162.7	163.2	162.0	160.2		1.72	1.72	1.72	1.71
1996		160.1	158.0	164.7	166.7		1.71	1.68	1.76	1.79
1997		162.6	163.0	159.2	158.2		1.75	1.75	1.72	1.71
1998		160.4	158.5	160.0	156.9		1.74	1.71	1.74	1.71
1999		156.4	157.2	154.7	153.7		1.70	1.71	1.69	1.68
2000		151.6	151.3	150.2	151.3		1.66	1.65	1.65	1.66
2001		150.0	149.2	148.1	147.4		1.65	1.64	1.63	1.63
2002		148.2	147.9	149.5	150.5		1.64	1.63	1.66	1.67
2003		152.6	156.1	157.0	155.7		1.69	1.73	1.75	1.73
2004		159.4	159.1	159.6	161.7		1.78	1.77	1.78	1.80
2005		160.1	161.1	163.5	161.2		1.79	1.79	1.82	1.80

Note: Figures may not add exactly due to rounding - see section 2.15.
1 See sections 2.1 and 2.11.
2 See section 3.1.

Table 2.2   Stillbirths: quarter of occurrence, 1995-2005   England and Wales

Year	Total	Quarter ending			
		31 March	30 June	30 September	31 December
1995	**3,600**	958	865	904	873
1996	**3,539**	875	886	917	861
1997	**3,439**	867	870	839	863
1998	**3,417**	865	849	824	879
1999	**3,305**	867	834	818	786
2000	**3,203**	808	776	797	822
2001	**3,159**	785	799	745	830
2002	**3,372**	827	869	849	827
2003	**3,612**	867	914	939	892
2004	**3,686**	901	916	991	878
2005	**3,483**	810	920	900	853

Table 2.3   Live births (actual and seasonally adjusted numbers and rates[1]):   **England and Wales**
month of occurrence, 1995-2005

Year	Total	January	February	March	April	May	June	July	August	September	October	November	December
Numbers (thousands)													
Live births													
1995	**648.1**	53.5	49.7	55.4	52.2	56.7	55.8	56.5	55.6	55.3	54.9	51.4	51.2
1996	**649.5**	53.5	50.4	53.4	50.5	53.8	53.7	57.6	56.0	56.4	56.2	53.8	54.2
1997	**643.1**	54.5	49.5	54.1	54.0	55.4	53.9	56.4	54.8	53.7	53.0	50.7	53.1
1998	**635.9**	53.4	48.8	53.6	52.4	53.0	53.1	56.4	54.4	55.3	53.6	50.1	51.6
1999	**621.9**	51.2	47.6	53.3	50.8	53.5	53.0	54.5	52.9	52.7	51.0	49.9	51.6
2000	**604.4**	50.5	47.0	51.1	49.0	51.8	49.9	52.6	51.9	50.4	50.9	49.8	49.5
2001	**594.6**	50.6	45.2	49.8	47.8	51.5	49.5	51.3	51.0	50.7	51.4	48.4	47.6
2002	**596.1**	49.2	45.2	49.0	47.9	50.7	48.5	52.0	51.1	51.9	52.2	48.6	49.7
2003	**621.5**	50.6	46.0	50.8	50.8	52.9	51.5	55.4	53.3	54.1	54.2	50.6	51.3
2004	**639.7**	53.0	49.4	52.8	51.4	52.7	53.3	55.6	54.3	55.6	55.4	52.9	53.4
2005	**645.8**	52.5	48.1	53.7	52.1	53.6	54.1	56.5	57.2	56.4	55.6	52.3	53.8
Seasonally adjusted live births[2]													
1995		54.2	54.4	54.2	53.9	54.7	54.7	54.1	54.1	53.8	54.5	53.2	52.5
1996		53.6	53.4	53.1	52.0	52.5	53.5	54.5	54.9	55.3	55.5	55.9	55.4
1997		54.5	54.3	53.8	55.1	54.4	53.6	53.2	53.8	52.2	52.1	52.8	53.3
1998		53.6	53.5	53.3	53.3	52.4	52.8	53.2	53.3	53.6	53.0	51.9	51.9
1999		51.7	52.3	52.4	52.0	52.6	52.6	51.9	51.8	51.0	50.6	51.2	51.9
2000		51.0	50.1	50.5	50.8	50.4	50.1	50.6	50.4	49.2	50.2	50.7	50.4
2001		50.4	49.9	49.7	49.4	50.0	49.8	49.3	49.3	49.4	49.7	49.4	48.3
2002		49.1	49.8	49.3	49.3	49.4	49.3	49.3	49.8	50.4	50.4	49.6	50.5
2003		50.5	50.8	51.2	51.9	52.1	52.1	52.6	52.3	52.2	52.1	52.1	51.5
2004		53.5	53.1	52.8	52.9	52.6	53.5	53.1	53.2	53.3	53.9	54.1	53.7
2005		53.3	53.3	53.6	53.6	53.5	54.0	54.1	55.1	54.3	53.9	53.1	54.2
Total fertility rates (TFR)													
Actual TFR													
1995	**1.72**	1.67	1.71	1.72	1.68	1.76	1.80	1.76	1.74	1.79	1.72	1.67	1.61
1996	**1.74**	1.69	1.70	1.68	1.64	1.69	1.75	1.82	1.77	1.84	1.78	1.76	1.72
1997	**1.73**	1.73	1.74	1.71	1.76	1.75	1.76	1.79	1.74	1.76	1.68	1.67	1.69
1998	**1.72**	1.70	1.72	1.71	1.72	1.69	1.75	1.80	1.74	1.83	1.72	1.66	1.66
1999	**1.70**	1.64	1.69	1.71	1.68	1.71	1.75	1.75	1.70	1.75	1.64	1.66	1.66
2000	**1.65**	1.63	1.62	1.65	1.63	1.67	1.66	1.70	1.68	1.69	1.65	1.67	1.61
2001	**1.63**	1.64	1.62	1.61	1.59	1.66	1.65	1.66	1.65	1.70	1.67	1.63	1.55
2002	**1.65**	1.60	1.62	1.59	1.61	1.65	1.63	1.69	1.67	1.75	1.70	1.64	1.63
2003	**1.73**	1.65	1.67	1.66	1.71	1.73	1.74	1.81	1.75	1.83	1.77	1.71	1.68
2004	**1.78**	1.74	1.74	1.73	1.75	1.73	1.81	1.83	1.79	1.89	1.82	1.80	1.76
2005	**1.80**	1.72	1.75	1.76	1.76	1.76	1.83	1.86	1.88	1.92	1.82	1.77	1.76
Seasonally adjusted TFR[2]													
1995		1.72	1.73	1.72	1.71	1.73	1.74	1.72	1.72	1.72	1.74	1.70	1.68
1996		1.72	1.71	1.70	1.66	1.68	1.71	1.75	1.76	1.78	1.79	1.80	1.79
1997		1.76	1.75	1.73	1.78	1.75	1.73	1.72	1.74	1.69	1.69	1.71	1.73
1998		1.74	1.74	1.73	1.73	1.70	1.71	1.73	1.74	1.75	1.73	1.70	1.70
1999		1.69	1.71	1.71	1.69	1.71	1.72	1.70	1.70	1.67	1.66	1.68	1.71
2000		1.67	1.64	1.65	1.66	1.65	1.64	1.66	1.66	1.62	1.65	1.67	1.66
2001		1.66	1.64	1.63	1.63	1.64	1.64	1.62	1.63	1.63	1.64	1.64	1.60
2002		1.63	1.65	1.63	1.63	1.63	1.63	1.64	1.66	1.67	1.67	1.65	1.68
2003		1.69	1.69	1.71	1.73	1.73	1.74	1.75	1.74	1.74	1.74	1.74	1.72
2004		1.79	1.78	1.76	1.77	1.76	1.79	1.78	1.78	1.79	1.81	1.81	1.79
2005		1.78	1.78	1.79	1.79	1.79	1.80	1.81	1.84	1.82	1.80	1.77	1.81

Note: Figures may not add exactly due to rounding - see section 2.15.
1 See sections 2.1 and 2.11.
2 See section 3.1.

Table 2.4 Maternities, live births and stillbirths: quarter and month of occurrence[1], within/outside marriage and sex, 2005 — England and Wales

Quarter/month of occurrence	Maternities	Live births			Stillbirths			Live births		Stillbirths	
		Total	Within marriage	Outside marriage	Total	Within marriage	Outside marriage	Male	Female	Male	Female
Annual Total	**639,627**	**645,835**	**369,330**	**276,505**	**3,483**	**1,861**	**1,622**	**330,600**	**315,235**	**1,796**	**1,687**
March quarter	152,781	**154,278**	87,993	66,285	**810**	448	362	79,039	75,239	423	387
June quarter	158,312	**159,752**	93,157	66,595	**920**	461	459	81,727	78,025	474	446
September quarter	168,562	**170,152**	96,440	73,712	**900**	489	411	87,039	83,113	469	431
December quarter	159,972	**161,653**	91,740	69,913	**853**	463	390	82,795	78,858	430	423
January	51,968	**52,527**	29,534	22,993	**271**	154	117	26,886	25,641	141	130
February	47,691	**48,080**	27,363	20,717	**293**	167	126	24,650	23,430	161	132
March	53,122	**53,671**	31,096	22,575	**246**	127	119	27,503	26,168	121	125
April	51,575	**52,064**	30,391	21,673	**279**	150	129	26,761	25,303	145	134
May	53,206	**53,610**	31,389	22,221	**322**	164	158	27,327	26,283	153	169
June	53,531	**54,078**	31,377	22,701	**319**	147	172	27,639	26,439	176	143
July	56,018	**56,526**	32,096	24,430	**297**	174	123	28,928	27,598	150	147
August	56,691	**57,230**	32,489	24,741	**321**	173	148	29,256	27,974	165	156
September	55,853	**56,396**	31,855	24,541	**282**	142	140	28,855	27,541	154	128
October	55,055	**55,614**	32,012	23,602	**308**	174	134	28,468	27,146	161	147
November	51,678	**52,279**	29,662	22,617	**254**	136	118	26,703	25,576	133	121
December	53,239	**53,760**	30,066	23,694	**291**	153	138	27,624	26,136	136	155

1 Including a small number of births which occurred in 2004 and where registered after the 'cut off' date in 2005 - see section 2.2.

Table 2.5 Live birth occurrences[1] in 2005: quarter and month of occurrence and month of registration — England and Wales

Quarter/month of occurrence	Month of registration[1]												Registrations from 1.1.06 to 25.2.06	Total
	January	February	March	April	May	June	July	August	September	October	November	December		
Annual Total	**23,510**	**45,546**	**51,574**	**55,636**	**50,592**	**57,140**	**53,927**	**55,872**	**60,081**	**54,526**	**54,745**	**44,889**	**37,797**	**645,835**
March	23,510	45,519	51,376	31,118	2,533	174	24	5	9	5	4	1	-	**154,278**
June	-	-	1	24,485	48,050	56,958	27,820	2,315	84	22	7	6	4	**159,752**
September	-	1	5	3	-	2	26,079	53,548	59,987	28,045	2,344	105	33	**170,152**
December	-	26	192	30	9	6	4	4	1	26,454	52,390	44,777	37,760	**161,653**
January	23,510	25,652	3,199	128	18	10	3	1	4	1	1	-	-	**52,527**
February	-	19,867	25,315	2,799	85	7	4	1	1	-	1	-	-	**48,080**
March	-	-	22,862	28,191	2,430	157	17	3	4	4	2	1	-	**53,671**
April	-	-	-	24,482	24,814	2,619	127	14	6	1	-	-	1	**52,064**
May	-	-	1	1	23,235	27,933	2,334	84	10	6	1	4	1	**53,610**
June	-	-	-	2	1	26,406	25,359	2,217	68	15	6	2	2	**54,078**
July	-	-	-	1	-	1	26,078	28,027	2,315	75	19	8	2	**56,526**
August	-	1	2	1	-	-	1	25,519	29,610	1,992	89	10	5	**57,230**
September	-	-	3	1	-	1	-	2	28,062	25,978	2,236	87	26	**56,396**
October	-	-	13	7	1	1	1	2	-	26,452	26,993	1,977	167	**55,614**
November	-	3	20	4	3	2	1	1	1	2	25,396	23,753	3,093	**52,279**
December	-	23	159	19	5	3	2	1	-	-	1	19,047	34,500	**53,760**

1 Including a small number of births which occurred in 2004 and were registered after the 'cut-off' date in 2005 - see section 2.2.

Table 3.1   Live births: age of mother and occurrence within/outside marriage, 1995-2005   **England and Wales**
a. numbers

Year	Age of mother at birth							
	All ages	Under 20	20-24	25-29	30-34	35-39	40-44	45 and over
	All live births							
1995	**648,138**	41,938	130,744	217,418	181,202	65,517	10,779	540
1996	**649,485**	44,667	125,732	211,103	186,377	69,503	11,516	587
1997	**643,095**	46,372	118,589	202,792	187,528	74,900	12,332	582
1998	**635,901**	48,285	113,537	193,144	188,499	78,881	12,980	575
1999	**621,872**	48,375	110,722	181,931	185,311	81,281	13,617	635
2000	**604,441**	45,846	107,741	170,701	180,113	84,974	14,403	663
2001	**594,634**	44,189	108,844	159,926	178,920	86,495	15,499	761
2002	**596,122**	43,467	110,959	153,379	180,532	90,449	16,441	895
2003	**621,469**	44,236	116,622	156,931	187,214	97,386	18,205	875
2004	**639,721**	45,094	121,072	159,984	190,550	102,228	19,884	909
2005	**645,835**	44,830	122,145	164,348	188,153	104,113	21,155	1,091
	Live births within marriage							
1995	**428,189**	5,623	61,029	157,855	144,200	51,129	7,944	409
1996	**416,822**	5,365	54,651	148,770	145,898	53,265	8,421	452
1997	**404,873**	5,233	49,068	139,383	145,293	56,671	8,797	428
1998	**395,290**	5,278	45,724	130,747	144,599	59,320	9,189	433
1999	**379,983**	5,333	43,190	120,716	140,330	60,470	9,466	478
2000	**365,836**	4,742	40,262	111,606	136,165	62,671	9,910	480
2001	**356,548**	4,640	40,736	103,131	133,710	63,202	10,582	547
2002	**354,090**	4,582	40,712	97,583	134,093	65,369	11,110	641
2003	**364,244**	4,338	40,887	98,694	138,002	69,595	12,113	615
2004	**369,997**	4,063	41,285	98,539	139,838	72,555	13,098	619
2005	**369,330**	3,654	40,010	99,957	137,401	73,810	13,760	738
	Live births outside marriage							
1995	**219,949**	36,315	69,715	59,563	37,002	14,388	2,835	131
1996	**232,663**	39,302	71,081	62,333	40,479	16,238	3,095	135
1997	**238,222**	41,139	69,521	63,409	42,235	18,229	3,535	154
1998	**240,611**	43,007	67,813	62,397	43,900	19,561	3,791	142
1999	**241,889**	43,042	67,532	61,215	44,981	20,811	4,151	157
2000	**238,605**	41,104	67,479	59,095	43,948	22,303	4,493	183
2001	**238,086**	39,549	68,108	56,795	45,210	23,293	4,917	214
2002	**242,032**	38,885	70,247	55,796	46,439	25,080	5,331	254
2003	**257,225**	39,898	75,735	58,237	49,212	27,791	6,092	260
2004	**269,724**	41,031	79,787	61,445	50,712	29,673	6,786	290
2005	**276,505**	41,176	82,135	64,391	50,752	30,303	7,395	353

Table 3.1   Live births: age of mother and occurrence within/outside marriage, 1995-2005   **England and Wales**
            b. rates[1]

Year	Age of mother at birth							
	All ages	Under 20	20-24	25-29	30-34	35-39	40-44	45 and over
	All live births per 1,000 women[2]							
1995	**60.5**	28.5	76.4	108.4	88.3	36.3	6.5	0.3
1996	**60.6**	29.7	77.0	106.6	89.8	37.5	6.9	0.3
1997	**60.0**	30.2	76.0	104.3	89.8	39.4	7.3	0.3
1998	**59.2**	30.9	74.9	101.5	90.6	40.4	7.5	0.3
1999	**57.8**	30.9	73.0	98.3	89.6	40.6	7.7	0.4
2000	**55.9**	29.3	70.0	94.3	87.9	41.4	8.0	0.4
2001	**54.7**	28.0	69.0	91.7	88.0	41.5	8.4	0.5
2002	**54.7**	27.0	69.2	91.6	89.8	43.0	8.6	0.5
2003	**56.8**	26.8	71.2	96.4	94.8	46.4	9.3	0.5
2004	**58.2**	26.9	72.7	98.4	99.4	48.9	9.9	0.5
2005	**58.4**	26.3	71.7	98.8	100.9	50.3	10.3	0.6
	Live births within marriage per 1,000 married women[2]							
1995	**82.7**	270.7	203.4	164.0	108.4	39.2	6.3	0.3
1996	**82.2**	258.7	210.5	164.2	110.9	40.4	6.7	0.3
1997	**81.6**	257.7	218.2	165.1	112.4	42.6	7.0	0.3
1998	**81.3**	260.4	227.4	167.0	114.8	44.2	7.3	0.3
1999	**79.6**	269.9	230.0	166.5	114.8	44.7	7.5	0.4
2000	**77.9**	268.8	224.2	164.8	115.2	46.0	7.8	0.4
2001	**77.3**	287.1	228.9	164.9	117.0	46.6	8.2	0.5
2002	**78.8**	352.4	245.1	172.0	122.5	48.7	8.5	0.5
2003	**83.3**	370.2	254.3	188.5	132.4	52.9	9.2	0.5
2004	**86.8**	382.6	264.0	198.4	142.0	56.5	9.9	0.5
2005	**88.8**	411.0	269.1	206.8	147.5	59.2	10.3	0.6
	Live births outside marriage per 1,000 single, widowed and divorced women[2]							
1995	**39.7**	25.0	49.4	57.2	51.3	28.9	6.9	0.3
1996	**41.2**	26.5	51.8	58.1	53.3	30.4	7.3	0.3
1997	**41.3**	27.2	52.0	57.6	53.1	31.9	7.9	0.4
1998	**40.9**	27.9	51.5	55.7	53.5	32.1	8.0	0.3
1999	**40.4**	27.8	50.8	54.4	53.1	32.0	8.4	0.4
2000	**39.0**	26.6	49.6	52.2	50.7	32.3	8.5	0.4
2001	**38.0**	25.3	48.7	50.7	50.7	32.1	8.8	0.5
2002	**37.7**	24.4	48.8	50.4	50.7	33.0	8.9	0.5
2003	**39.1**	24.4	51.3	52.8	52.8	35.3	9.6	0.5
2004	**40.1**	24.6	52.9	54.4	54.4	36.7	10.0	0.5
2005	**40.1**	24.3	52.9	54.6	54.4	36.8	10.3	0.6

1 The rates for women of all ages, under 20 and 45 and over are based upon the population of women aged 15-44, 15-19 and 45-49 respectively.
2 See sections 2.1 and 2.11.

Table 3.2   Maternities (total), live births and stillbirths (total and female):
age of mother and occurrence within/outside marriage, 2005

Age of mother at birth	Maternities			Births			
	Total	Within marriage	Outside marriage	Live births		Stillbirths	
				Total	Female	Total	Female
All ages	639,627	365,089	274,538	645,835	315,235	3,483	1,687
11	4	-	4	4	-	-	-
12	4	-	4	4	3	-	-
13	15	-	15	15	10	-	-
14	188	5	183	189	91	-	-
15	977	10	967	977	505	6	2
16	3,511	67	3,444	3,514	1,710	20	8
17	8,694	313	8,381	8,685	4,201	51	26
18	13,380	950	12,430	13,381	6,549	87	45
19	18,040	2,311	15,729	18,061	8,924	122	49
Under 20	44,813	3,656	41,157	44,830	21,993	286	130
20	20,890	3,952	16,938	20,924	10,223	127	54
21	22,289	5,668	16,621	22,376	10,981	121	60
22	24,196	7,691	16,505	24,260	11,896	143	76
23	26,027	9,999	16,028	26,154	12,809	138	68
24	28,242	12,552	15,690	28,431	13,764	110	53
20-24	121,644	39,862	81,782	122,145	59,673	639	311
25	30,562	15,450	15,112	30,776	15,070	152	71
26	31,420	17,736	13,684	31,636	15,488	176	77
27	31,649	19,441	12,208	31,913	15,653	195	98
28	33,255	21,866	11,389	33,557	16,240	171	89
29	36,100	24,627	11,473	36,466	17,851	179	85
25-29	162,986	99,120	63,866	164,348	80,302	873	420
30	37,526	26,743	10,783	37,931	18,609	178	89
31	39,084	28,348	10,736	39,584	19,204	170	83
32	37,937	27,981	9,956	38,403	18,688	188	99
33	36,818	27,195	9,623	37,332	18,023	175	91
34	34,384	25,304	9,080	34,903	17,155	197	98
30-34	185,749	135,571	50,178	188,153	91,679	908	460
35	29,469	21,450	8,019	29,936	14,658	147	62
36	24,873	17,681	7,192	25,265	12,445	135	75
37	20,162	14,208	5,954	20,507	9,836	116	52
38	15,731	10,913	4,818	15,943	7,676	115	48
39	12,301	8,370	3,931	12,462	6,024	97	44
35-39	102,536	72,622	29,914	104,113	50,639	610	281
40	8,640	5,743	2,897	8,794	4,350	58	32
41	5,666	3,671	1,995	5,749	2,761	38	18
42	3,497	2,200	1,297	3,521	1,752	32	14
43	1,982	1,264	718	2,010	1,023	14	7
44	1,070	674	396	1,081	546	11	6
40-44	20,855	13,552	7,303	21,155	10,432	153	77
45	560	370	190	569	256	7	4
46	215	138	77	227	113	3	2
47	108	83	25	112	57	2	1
48	60	43	17	69	31	-	-
49	44	35	9	47	29	2	1
45-49	987	669	318	1,024	486	14	8
50 and over	57	37	20	67	31	-	-

England and Wales

| Births within marriage | | | | Births outside marriage | | | | Age of mother at birth |
| Live births | | Stillbirths | | Live births | | Stillbirths | | |
Total	Female	Total	Female	Total	Female	Total	Female	
369,330	**179,942**	**1,861**	**908**	**276,505**	**135,293**	**1,622**	**779**	**All ages**
-	-	-	-	4	-	-	-	11
-	-	-	-	4	3	-	-	12
-	-	-	-	15	10	-	-	13
5	3	-	-	184	88	-	-	14
10	5	-	-	967	500	6	2	15
66	31	1	1	3,448	1,679	19	7	16
310	122	4	1	8,375	4,079	47	25	17
946	457	8	6	12,435	6,092	79	39	18
2,317	1,122	13	6	15,744	7,802	109	43	19
3,654	**1,740**	**26**	**14**	**41,176**	**20,253**	**260**	**116**	**Under 20**
3,947	1,909	37	16	16,977	8,314	90	38	20
5,678	2,752	35	12	16,698	8,229	86	48	21
7,695	3,761	54	32	16,565	8,135	89	44	22
10,042	4,938	61	34	16,112	7,871	77	34	23
12,648	6,153	36	16	15,783	7,611	74	37	24
40,010	**19,513**	**223**	**110**	**82,135**	**40,160**	**416**	**201**	**20-24**
15,555	7,629	70	34	15,221	7,441	82	37	25
17,847	8,773	100	42	13,789	6,715	76	35	26
19,610	9,581	119	61	12,303	6,072	76	37	27
22,051	10,679	112	59	11,506	5,561	59	30	28
24,894	12,167	110	52	11,572	5,684	69	33	29
99,957	**48,829**	**511**	**248**	**64,391**	**31,473**	**362**	**172**	**25-29**
27,025	13,300	116	58	10,906	5,309	62	31	30
28,743	13,914	104	55	10,841	5,290	66	28	31
28,332	13,782	130	69	10,071	4,906	58	30	32
27,589	13,309	115	58	9,743	4,714	60	33	33
25,712	12,628	144	68	9,191	4,527	53	30	34
137,401	**66,933**	**609**	**308**	**50,752**	**24,746**	**299**	**152**	**30-34**
21,809	10,634	101	42	8,127	4,024	46	20	35
17,980	8,885	90	47	7,285	3,560	45	28	36
14,460	6,902	69	36	6,047	2,934	47	16	37
11,080	5,299	66	23	4,863	2,377	49	25	38
8,481	4,124	58	28	3,981	1,900	39	16	39
73,810	**35,844**	**384**	**176**	**30,303**	**14,795**	**226**	**105**	**35-39**
5,856	2,869	33	18	2,938	1,481	25	14	40
3,726	1,777	26	12	2,023	984	12	6	41
2,212	1,069	25	9	1,309	683	7	5	42
1,286	659	10	5	724	364	4	2	43
680	359	6	3	401	187	5	3	44
13,760	**6,733**	**100**	**47**	**7,395**	**3,699**	**53**	**30**	**40-44**
378	171	3	3	191	85	4	1	45
147	74	1	-	80	39	2	2	46
85	43	2	1	27	14	-	-	47
50	20	-	-	19	11	-	-	48
37	22	2	1	10	7	-	-	49
697	**330**	**8**	**5**	**327**	**156**	**6**	**3**	**45-49**
41	20	-	-	26	11	-	-	50 and over

Table 3.3 Live births within marriage (numbers and rates[1]): age of father, 1995-2005   **England and Wales**

Year	Age of father at birth											
	All ages	Under 20	20-24	25-29	30-34	35-39	40-44	45-49	50-54	55-59	60-64	65 and over
	Numbers											
1995	**428,189**	1,015	28,873	120,070	157,518	80,946	26,934	8,826	2,475	1,008	364	160
1996	**416,822**	867	25,013	112,012	155,022	83,279	27,819	8,775	2,584	933	347	171
1997	**404,873**	940	21,895	103,320	151,162	86,097	28,579	8,684	2,794	894	350	158
1998	**395,290**	964	19,755	96,415	148,267	87,684	29,345	8,701	2,807	820	368	164
1999	**379,983**	1,028	18,703	87,896	141,629	88,329	29,986	8,328	2,756	878	304	146
2000	**365,836**	907	16,762	79,351	135,810	89,607	30,782	8,446	2,861	843	330	137
2001	**356,548**	907	16,872	72,668	132,570	88,657	31,981	8,756	2,775	863	344	155
2002	**354,090**	907	16,827	68,390	130,835	90,871	33,337	8,858	2,676	920	320	149
2003	**364,244**	835	16,753	66,854	133,951	95,381	36,257	9,858	2,839	1,002	313	201
2004	**369,997**	753	16,743	66,666	134,002	98,579	38,324	10,532	2,817	1,051	323	207
2005	**369,330**	695	16,028	65,868	131,463	99,486	40,179	11,100	2,957	1,047	313	194
	Rates: live births within marriage per 1,000 married men by age											
1995	**37.7**	169.1	210.8	171.7	136.9	66.6	22.3	6.4	2.1	0.9	0.4	0.1
1996	**36.9**	145.2	214.6	172.3	136.6	67.8	23.2	6.4	2.2	0.9	0.4	0.1
1997	**36.0**	155.1	222.1	173.6	136.1	69.7	23.9	6.7	2.2	0.9	0.4	0.1
1998	**35.3**	155.8	229.7	177.2	137.5	70.7	24.4	7.0	2.1	0.8	0.4	0.1
1999	**34.1**	164.4	238.3	176.8	135.7	70.7	24.8	6.9	2.0	0.8	0.3	0.1
2000	**32.9**	153.3	225.2	173.0	134.9	71.4	25.1	7.1	2.1	0.8	0.3	0.1
2001	**32.2**	165.1	227.0	173.1	136.5	70.9	25.7	7.4	2.1	0.7	0.4	0.1
2002	**32.1**	206.6	242.3	184.1	142.0	73.7	26.6	7.5	2.1	0.7	0.3	0.1
2003	**33.3**	221.0	244.4	198.1	154.3	79.0	28.7	8.4	2.4	0.8	0.3	0.1
2004	**34.1**	258.3	248.2	209.6	165.3	83.7	30.2	8.9	2.4	0.8	0.3	0.1
2005	**34.2**	335.4	246.4	216.2	173.4	87.0	31.6	9.3	2.6	0.8	0.3	0.1

1 See sections 2.1 and 2.11.

Table 3.4 Paternities (total), live births and stillbirths (total and male): England and Wales
age of father and occurrence within/outside marriage, 2005

Age of father at birth	Paternities within marriage	Births within marriage				Births outside marriage registered[1] on joint information of parents			
		Live births		Stillbirths		Live births		Stillbirths	
		Total	Male	Total	Male	Total	Male	Total	Male
All ages	365,089	369,330	189,388	1,861	953	231,335	118,359	1,324	674
13	-	-	-	-	-	4	1	-	-
14	1	1	1	-	-	18	10	-	-
15	3	3	1	-	-	149	78	2	-
16	7	7	2	-	-	608	316	2	1
17	47	47	26	-	-	1,791	914	12	6
18	135	133	69	2	1	3,574	1,812	22	13
19	507	504	250	7	4	6,080	3,071	30	15
Under 20	700	695	349	9	5	12,224	6,202	68	35
20	1,030	1,027	536	12	6	8,078	4,106	58	33
21	1,790	1,806	902	9	4	9,342	4,722	58	30
22	2,839	2,844	1,459	14	10	10,576	5,440	55	24
23	4,260	4,269	2,211	26	14	11,133	5,836	73	40
24	6,046	6,082	3,116	34	20	11,849	5,973	65	29
20-24	15,965	16,028	8,224	95	54	50,978	26,077	309	156
25	8,234	8,286	4,311	54	26	12,153	6,267	74	39
26	10,274	10,333	5,267	58	30	11,910	6,073	61	30
27	12,615	12,675	6,481	75	39	10,849	5,581	45	23
28	15,554	15,682	8,056	59	29	10,599	5,440	56	27
29	18,738	18,892	9,781	95	50	10,868	5,584	58	31
25-29	65,415	65,868	33,896	341	174	56,379	28,945	294	150
30	22,613	22,844	11,664	93	37	10,778	5,554	57	31
31	25,166	25,440	12,993	114	61	10,662	5,456	57	26
32	26,983	27,301	14,122	115	68	10,492	5,426	56	20
33	27,890	28,212	14,441	143	84	10,091	5,089	48	28
34	27,296	27,666	14,154	115	54	9,666	4,882	57	34
30-34	129,948	131,463	67,374	580	304	51,689	26,407	275	139
35	24,952	25,274	12,829	121	63	8,769	4,572	38	19
36	22,598	22,938	11,779	97	42	8,121	4,178	52	25
37	19,682	19,953	10,209	106	45	7,053	3,563	52	32
38	16,645	16,869	8,536	103	47	6,256	3,176	43	23
39	14,266	14,452	7,462	72	40	5,604	2,899	40	21
35-39	98,143	99,486	50,815	499	237	35,803	18,388	225	120
40	12,024	12,233	6,241	47	22	4,746	2,369	27	11
41	9,497	9,627	4,956	59	34	3,993	2,024	22	14
42	7,503	7,595	3,912	52	26	3,172	1,610	20	13
43	5,903	5,989	3,125	40	22	2,605	1,352	10	3
44	4,650	4,735	2,448	20	13	2,101	1,059	19	10
40-44	39,577	40,179	20,682	218	117	16,617	8,414	98	51
45	3,554	3,604	1,882	30	18	1,611	854	14	8
46	2,617	2,671	1,362	17	12	1,279	644	8	3
47	2,040	2,065	1,039	22	10	986	504	4	1
48	1,540	1,572	836	5	2	785	402	4	2
49	1,165	1,188	632	4	1	629	327	9	3
45-49	10,916	11,100	5,751	78	43	5,290	2,731	39	17
50-54	2,911	2,957	1,502	27	11	1,567	796	11	5
55-59	1,026	1,047	526	12	8	573	275	3	1
60-64	298	313	151	1	-	144	83	1	-
65-69	133	135	83	-	-	55	31	1	-
70-74	35	35	21	-	-	12	8	-	-
75 and over	22	24	14	1	-	4	2	-	-

1 45,170 live births and 298 stillbirths occurred outside marriage and were registered by the mother alone. For these sole registrations only the mother's details were recorded.

Table 3.5 Rates[1] of paternities within marriage by age of father, and rates[1] of live births and stillbirths within marriage by age of father, 2005

England and Wales

Age of father at birth	Paternities within marriage per 1,000 married men	Births within marriage per 1,000 married men			
		Live births		Stillbirths	
		Total	Male	Total	Male
All ages	**33.8**	**34.2**	**17.54**	**0.17**	**0.09**
Under 20	337.8	335.4	168.44	*4.34*	*2.41*
20-24	245.4	246.4	126.44	1.46	0.83
25-29	214.7	216.2	111.25	1.12	0.57
30-34	171.4	173.4	88.88	0.77	0.40
35-39	85.8	87.0	44.42	0.44	0.21
40-44	31.1	31.6	16.27	0.17	0.09
45-49	9.1	9.3	4.82	0.07	0.04
50-54	2.6	2.6	1.32	0.02	*0.01*
55-59	0.8	0.8	0.41	*0.01*	*0.01*
60-64	0.3	0.3	0.15	*0.00*	-
65 and over	0.1	0.1	0.05	*0.00*	-

1 See sections 2.1 and 2.11.

Table 3.6 Live births within marriage: age of mother and of father, 2005

England and Wales

Age of father at birth	Age of mother at birth								
	All ages	Under 20	20-24	25-29	30-34	35-39	40-44	45-49	50 and over
All ages	369,330	3,654	40,010	99,957	137,401	73,810	13,760	697	41
Under 20	695	346	276	47	19	7	-	-	-
20-24	16,028	1,636	10,118	3,281	752	204	37	-	-
25-29	65,868	1,041	17,169	35,045	10,489	1,866	247	10	1
30-34	131,463	465	8,530	41,453	66,624	13,155	1,196	36	4
35-39	99,486	116	2,778	14,527	43,546	34,989	3,418	106	6
40-44	40,179	34	792	4,001	12,049	17,541	5,586	173	3
45-49	11,100	13	228	1,094	2,838	4,371	2,326	222	8
50-54	2,957	2	67	313	709	1,131	640	90	5
55-59	1,047	-	32	127	250	383	215	33	7
60-64	313	-	11	38	76	101	59	22	6
65 and over	194	1	9	31	49	62	36	5	1

Table 3.7 Stillbirths within marriage: age of mother and of father, 2005

England and Wales

Age of father at birth	Age of mother at birth								
	All ages	Under 20	20-24	25-29	30-34	35-39	40-44	45-49	50 and over
All ages	1,861	26	223	511	609	384	100	8	-
Under 20	9	3	6	-	-	-	-	-	-
20-24	95	10	58	21	5	1	-	-	-
25-29	341	8	93	181	51	7	1	-	-
30-34	580	4	44	204	266	56	6	-	-
35-39	499	1	16	75	200	181	25	1	-
40-44	218	-	4	21	55	96	39	3	-
45-49	78	-	-	5	21	27	23	2	-
50-54	27	-	2	4	9	10	2	-	-
55 and over	14	-	-	-	2	6	4	2	-

Table 3.8 Live births outside marriage: age of mother and of father and whether sole or joint registration, 2005

England and Wales

Age of father at birth	Age of mother at birth							
	All ages	Under 20	20-24	25-29	30-34	35-39	40-44	45 and over
	Total live births outside marriage							
	276,505	**41,176**	**82,135**	**64,391**	**50,752**	**30,303**	**7,395**	**353**
	Sole registration							
	45,170	10,752	14,471	8,906	6,187	3,749	1,051	54
	Joint registration							
All ages	**231,335**	**30,424**	**67,664**	**55,485**	**44,565**	**26,554**	**6,344**	**299**
Under 20	**12,224**	9,128	2,523	399	124	47	3	-
20-24	**50,978**	14,832	27,785	6,070	1,678	529	83	1
25-29	**56,379**	4,181	22,546	20,482	6,808	2,088	261	13
30-34	**51,689**	1,423	9,352	17,142	17,077	5,802	864	29
35-39	**35,803**	566	3,648	7,670	12,318	9,848	1,700	53
40-44	**16,617**	201	1,212	2,655	4,714	5,623	2,116	96
45-49	**5,290**	62	433	772	1,296	1,792	874	61
50-54	**1,567**	23	100	198	358	570	293	25
55-59	**573**	5	43	68	139	191	111	16
60 and over	**215**	3	22	29	53	64	39	5

Note: Figures for jointly registered live births outside marriage include a small number of cases registered by the mother alone, for which the father's name was included in the birth register.

Table 3.9   Live births outside marriage (numbers and percentages):
age of mother and whether sole or joint registration, 1995-2005

England and Wales

Year		All ages	Under 20	20-24	25-29	30-34	35 and over	All ages	Under 20	20-24	25-29	30-34	35 and over
		Numbers						**Percentage by age**					
1995	Total	219,949	36,315	69,715	59,563	37,002	17,354						
	Sole	47,916	11,880	15,626	10,864	6,485	3,061	21.8	32.7	22.4	18.2	17.5	17.6
	Joint	172,033	24,435	54,089	48,699	30,517	14,293	78.2	67.3	77.6	81.8	82.5	82.4
1996	Total	232,663	39,302	71,081	62,333	40,479	19,468						
	Sole	51,016	12,946	16,245	11,350	6,978	3,497	21.9	32.9	22.9	18.2	17.2	18.0
	Joint	181,647	26,356	54,836	50,983	33,501	15,971	78.1	67.1	77.1	81.8	82.8	82.0
1997	Total	238,222	41,139	69,521	63,409	42,235	21,918						
	Sole	50,582	13,223	15,590	11,015	6,997	3,757	21.2	32.1	22.4	17.4	16.6	17.1
	Joint	187,640	27,916	53,931	52,394	35,238	18,161	78.8	67.9	77.6	82.6	83.4	82.9
1998	Total	240,611	43,007	67,813	62,397	43,900	23,494						
	Sole	49,960	13,830	14,838	10,672	6,768	3,852	20.8	32.2	21.9	17.1	15.4	16.4
	Joint	190,651	29,177	52,975	51,725	37,132	19,642	79.2	67.8	78.1	82.9	84.6	83.6
1999	Total	241,889	43,042	67,532	61,215	44,981	25,119						
	Sole	48,203	13,194	14,477	9,936	6,615	3,981	19.9	30.7	21.4	16.2	14.7	15.8
	Joint	193,686	29,848	53,055	51,279	38,366	21,138	80.1	69.3	78.6	83.8	85.3	84.2
2000	Total	238,605	41,104	67,479	59,095	43,948	26,979						
	Sole	45,773	12,459	13,919	9,213	6,116	4,066	19.2	30.3	20.6	15.6	13.9	15.1
	Joint	192,832	28,645	53,560	49,882	37,832	22,913	80.8	69.7	79.4	84.4	86.1	84.9
2001	Total	238,086	39,549	68,108	56,795	45,210	28,424						
	Sole	43,744	11,669	13,583	8,582	5,875	4,035	18.4	29.5	19.9	15.1	13.0	14.2
	Joint	194,342	27,880	54,525	48,213	39,335	24,389	81.6	70.5	80.1	84.9	87.0	85.8
2002	Total	242,032	38,885	70,247	55,796	46,439	30,665						
	Sole	43,129	11,229	13,599	8,133	5,949	4,219	17.8	28.9	19.4	14.6	12.8	13.8
	Joint	198,903	27,656	56,648	47,663	40,490	26,446	82.2	71.1	80.6	85.4	87.2	86.2
2003	Total	257,225	39,898	75,735	58,237	49,212	34,143						
	Sole	44,875	11,192	14,434	8,555	6,094	4,600	17.4	28.1	19.1	14.7	12.4	13.5
	Joint	212,350	28,706	61,301	49,682	43,118	29,543	82.6	71.9	80.9	85.3	87.6	86.5
2004	Total	269,724	41,031	79,787	61,445	50,712	36,749						
	Sole	45,372	11,161	14,366	8,787	6,179	4,879	16.8	27.2	18.0	14.3	12.2	13.3
	Joint	224,352	29,870	65,421	52,658	44,533	31,870	83.2	72.8	82.0	85.7	87.8	86.7
2005	Total	276,505	41,176	82,135	64,391	50,752	38,051						
	Sole	45,170	10,752	14,471	8,906	6,187	4,854	16.3	26.1	17.6	13.8	12.2	12.8
	Joint	231,335	30,424	67,664	55,485	44,565	33,197	83.7	73.9	82.4	86.2	87.8	87.2

Note: Figures for jointly registered live births outside marriage include a small number of cases registered by the mother alone, for which the father's name was included in the birth register.

Table 3.10   Jointly registered live births outside marriage   **England and Wales**
             (numbers and percentages): age of mother and whether parents
             were usually resident at the same or different addresses, 1995-2005

Year	Addresses of mother and father	Age of mother at birth											
		All ages	Under 20	20-24	25-29	30-34	35 and over	**All ages**	Under 20	20-24	25-29	30-34	35 and over
		Numbers of jointly registered births outside marriage						Percentage by age					
1995	Total	**172,033**	**24,435**	**54,089**	**48,699**	**30,517**	**14,293**						
	Same	**127,789**	14,424	39,274	38,376	24,376	11,339	**74.3**	59.0	72.6	78.8	79.9	79.3
	Different	**44,244**	10,011	14,815	10,323	6,141	2,954	**25.7**	41.0	27.4	21.2	20.1	20.7
1996	Total	**181,647**	**26,356**	**54,836**	**50,983**	**33,501**	**15,971**						
	Same	**135,282**	15,410	39,978	40,384	26,875	12,635	**74.5**	58.5	72.9	79.2	80.2	79.1
	Different	**46,365**	10,946	14,858	10,599	6,626	3,336	**25.5**	41.5	27.1	20.8	19.8	20.9
1997	Total	**187,640**	**27,916**	**53,931**	**52,394**	**35,238**	**18,161**						
	Same	**141,740**	16,551	39,784	42,020	28,715	14,670	**75.5**	59.3	73.8	80.2	81.5	80.8
	Different	**45,900**	11,365	14,147	10,374	6,523	3,491	**24.5**	40.7	26.2	19.8	18.5	19.2
1998	Total	**190,651**	**29,177**	**52,975**	**51,725**	**37,132**	**19,642**						
	Same	**146,521**	17,589	39,818	42,212	30,938	15,964	**76.9**	60.3	75.2	81.6	83.3	81.3
	Different	**44,130**	11,588	13,157	9,513	6,194	3,678	**23.1**	39.7	24.8	18.4	16.7	18.7
1999	Total	**193,686**	**29,848**	**53,055**	**51,279**	**38,366**	**21,138**						
	Same	**149,584**	17,833	40,190	42,126	32,152	17,283	**77.2**	59.7	75.8	82.2	83.8	81.8
	Different	**44,102**	12,015	12,865	9,153	6,214	3,855	**22.8**	40.3	24.2	17.8	16.2	18.2
2000	Total	**192,832**	**28,645**	**53,560**	**49,882**	**37,832**	**22,913**						
	Same	**149,510**	17,011	40,450	41,236	31,795	19,018	**77.5**	59.4	75.5	82.7	84.0	83.0
	Different	**43,322**	11,634	13,110	8,646	6,037	3,895	**22.5**	40.6	24.5	17.3	16.0	17.0
2001	Total	**194,342**	**27,880**	**54,525**	**48,213**	**39,335**	**24,389**						
	Same	**150,421**	16,215	40,843	39,797	33,306	20,260	**77.4**	58.2	74.9	82.5	84.7	83.1
	Different	**43,921**	11,665	13,682	8,416	6,029	4,129	**22.6**	41.8	25.1	17.5	15.3	16.9
2002	Total	**198,903**	**27,656**	**56,648**	**47,663**	**40,490**	**26,446**						
	Same	**154,086**	15,824	42,315	39,554	34,326	22,067	**77.5**	57.2	74.7	83.0	84.8	83.4
	Different	**44,817**	11,832	14,333	8,109	6,164	4,379	**22.5**	42.8	25.3	17.0	15.2	16.6
2003	Total	**212,350**	**28,706**	**61,301**	**49,682**	**43,118**	**29,543**						
	Same	**163,374**	16,000	45,225	40,867	36,630	24,652	**76.9**	55.7	73.8	82.3	85.0	83.4
	Different	**48,976**	12,706	16,076	8,815	6,488	4,891	**23.1**	44.3	26.2	17.7	15.0	16.6
2004	Total	**224,352**	**29,870**	**65,421**	**52,658**	**44,533**	**31,870**						
	Same	**171,498**	16,262	47,793	43,038	37,820	26,585	**76.4**	54.4	73.1	81.7	84.9	83.4
	Different	**52,854**	13,608	17,628	9,620	6,713	5,285	**23.6**	45.6	26.9	18.3	15.1	16.6
2005	Total	**231,335**	**30,424**	**67,664**	**55,485**	**44,565**	**33,197**						
	Same	**175,555**	16,316	48,755	45,165	37,764	27,555	**75.9**	53.6	72.1	81.4	84.7	83.0
	Different	**55,780**	14,108	18,909	10,320	6,801	5,642	**24.1**	46.4	27.9	18.6	15.3	17.0

Note: Figures for jointly registered live births outside marriage include a small number of cases registered by the mother alone, for which the father's name
         was included in the birth register.

Table 4.1 Live births within marriage: number of previous live-born children and age of mother (five-year age groups), 1995-2005
  a. all married women

England and Wales

Year	Number of previous live-born children[1]						Year	Number of previous live-born children[1]					
	Total	0	1	2	3	4 or more		Total	0	1	2	3	4 or more
	All ages of mother at birth							**25-29**					
1995	**428,189**	168,118	158,102	66,692	22,320	12,957	1995	**157,855**	71,036	57,282	20,513	6,472	2,552
1996	**416,822**	163,020	153,780	65,291	22,038	12,693	1996	**148,770**	67,178	53,445	19,576	6,153	2,418
1997	**404,873**	157,049	150,404	63,220	21,515	12,685	1997	**139,383**	63,139	50,024	18,108	5,793	2,319
1998	**395,290**	155,708	146,850	60,391	20,327	12,014	1998	**130,747**	60,602	46,404	16,375	5,198	2,168
1999	**379,983**	153,423	139,490	56,375	19,517	11,178	1999	**120,716**	57,385	41,763	14,735	4,907	1,926
2000	**365,836**	146,509	134,691	54,920	18,590	11,126	2000	**111,606**	52,673	38,376	14,129	4,594	1,834
2001	**356,548**	143,908	132,228	52,111	17,553	10,748	2001	**103,131**	48,849	35,650	12,750	4,179	1,703
2002	**354,090**	145,241	130,346	50,307	17,560	10,636	2002	**97,583**	47,111	32,985	11,848	4,028	1,611
2003	**364,244**	150,982	132,873	51,981	17,641	10,767	2003	**98,694**	48,439	32,480	12,108	4,072	1,595
2004	**369,997**	154,548	133,671	52,508	18,297	10,973	2004	**98,539**	48,890	31,907	12,078	4,080	1,584
2005	**369,330**	155,953	132,007	52,198	18,097	11,075	2005	**99,957**	49,978	32,063	12,319	4,038	1,559
	Under 20							**30-34**					
1995	**5,623**	4,308	1,187	116	10	2	1995	**144,200**	46,573	58,458	26,098	8,335	4,736
1996	**5,365**	4,223	1,034	102	5	1	1996	**145,898**	47,742	59,084	25,968	8,426	4,678
1997	**5,233**	4,121	1,018	84	9	1	1997	**145,293**	48,090	59,414	25,112	8,116	4,561
1998	**5,278**	4,188	1,001	85	4	-	1998	**144,599**	49,529	58,944	23,998	7,740	4,388
1999	**5,333**	4,290	930	99	13	1	1999	**140,330**	50,031	56,622	22,314	7,293	4,070
2000	**4,742**	3,810	842	81	7	2	2000	**136,165**	49,400	54,760	21,120	6,832	4,053
2001	**4,640**	3,781	771	82	5	1	2001	**133,710**	49,695	53,780	19,828	6,456	3,951
2002	**4,582**	3,817	687	71	5	2	2002	**134,093**	51,026	53,730	18,999	6,538	3,800
2003	**4,338**	3,471	791	67	8	1	2003	**138,002**	54,170	54,343	19,248	6,413	3,828
2004	**4,063**	3,333	666	59	4	1	2004	**139,838**	55,534	54,479	19,315	6,604	3,906
2005	**3,654**	2,959	627	63	5	-	2005	**137,401**	55,734	52,763	18,682	6,388	3,834
	20-24							**35 and over**					
1995	**61,029**	32,340	20,634	6,453	1,367	235	1995	**59,482**	13,861	20,541	13,512	6,136	5,432
1996	**54,651**	28,925	18,481	5,789	1,231	225	1996	**62,138**	14,952	21,736	13,856	6,223	5,371
1997	**49,068**	25,891	16,558	5,257	1,153	209	1997	**65,896**	15,808	23,390	14,659	6,444	5,595
1998	**45,724**	24,303	15,537	4,679	993	212	1998	**68,942**	17,086	24,964	15,254	6,392	5,246
1999	**43,190**	23,474	14,443	4,175	910	188	1999	**70,414**	18,243	25,732	15,052	6,394	4,993
2000	**40,262**	21,571	13,744	3,950	827	170	2000	**73,061**	19,055	26,969	15,640	6,330	5,067
2001	**40,736**	22,188	13,723	3,897	751	177	2001	**74,331**	19,395	28,304	15,554	6,162	4,916
2002	**40,712**	22,432	13,506	3,866	739	169	2002	**77,120**	20,855	29,438	15,523	6,250	5,054
2003	**40,887**	22,187	13,915	3,834	777	174	2003	**82,323**	22,715	31,344	16,724	6,371	5,169
2004	**41,285**	22,551	13,806	3,994	791	143	2004	**86,272**	24,240	32,813	17,062	6,818	5,339
2005	**40,010**	22,131	13,203	3,757	770	149	2005	**88,308**	25,151	33,351	17,377	6,896	5,533

1 See section 2.9.

Table 4.1 Live births within marriage: number of previous live-born children and age of mother (five-year age groups), 1995-2005
b. women married once only

England and Wales

Year	Number of previous live-born children[1]						Year	Number of previous live-born children[1]					
	Total	0	1	2	3	4 or more		Total	0	1	2	3	4 or more
	All ages of mother at birth							**25-29**					
1995	**394,904**	158,963	147,266	59,295	18,705	10,675	1995	**150,644**	68,634	54,897	19,020	5,814	2,279
1996	**384,240**	153,994	143,223	58,094	18,546	10,383	1996	**142,375**	65,075	51,296	18,258	5,592	2,154
1997	**373,428**	148,419	140,155	56,425	18,023	10,406	1997	**133,555**	61,201	48,104	16,953	5,223	2,074
1998	**365,105**	147,207	136,867	53,797	17,292	9,942	1998	**125,642**	58,883	44,644	15,357	4,790	1,968
1999	**352,494**	145,500	130,434	50,644	16,645	9,271	1999	**116,462**	55,900	40,375	13,890	4,525	1,772
2000	**340,054**	139,210	126,117	49,451	15,999	9,277	2000	**107,887**	51,426	37,148	13,379	4,274	1,660
2001	**332,607**	136,975	124,162	47,145	15,248	9,077	2001	**99,996**	47,766	34,622	12,146	3,911	1,551
2002	**331,265**	138,490	122,629	45,760	15,323	9,063	2002	**94,901**	46,213	32,077	11,340	3,786	1,485
2003	**341,645**	144,274	125,276	47,325	15,478	9,292	2003	**96,292**	47,569	31,678	11,656	3,886	1,503
2004	**348,463**	147,907	126,502	48,280	16,237	9,537	2004	**96,334**	48,062	31,186	11,699	3,902	1,485
2005	**349,288**	149,852	125,473	48,183	16,181	9,599	2005	**97,845**	49,173	31,416	11,931	3,846	1,479
	Under 20							**30-34**					
1995	**5,609**	4,300	1,183	114	10	2	1995	**130,170**	42,557	53,710	23,095	6,923	3,885
1996	**5,355**	4,218	1,031	100	5	1	1996	**131,973**	43,639	54,427	23,009	7,039	3,859
1997	**5,226**	4,116	1,016	84	9	1	1997	**132,225**	44,292	54,982	22,343	6,803	3,805
1998	**5,273**	4,186	999	85	3	-	1998	**132,210**	45,773	54,755	21,354	6,634	3,694
1999	**5,329**	4,287	929	99	13	1	1999	**129,006**	46,439	52,704	20,112	6,286	3,465
2000	**4,737**	3,806	841	81	7	2	2000	**125,811**	46,204	51,177	18,990	5,960	3,480
2001	**4,638**	3,780	770	82	5	1	2001	**124,250**	46,644	50,523	17,963	5,669	3,451
2002	**4,574**	3,813	685	69	5	2	2002	**125,244**	48,079	50,631	17,381	5,796	3,357
2003	**4,331**	3,468	788	66	8	1	2003	**129,567**	51,362	51,430	17,638	5,737	3,400
2004	**4,060**	3,333	664	58	4	1	2004	**132,094**	52,921	51,795	17,936	5,937	3,505
2005	**3,649**	2,956	625	63	5	-	2005	**130,556**	53,432	50,443	17,421	5,820	3,440
	20-24							**35 and over**					
1995	**60,190**	32,056	20,331	6,278	1,309	216	1995	**48,291**	11,416	17,145	10,788	4,649	4,293
1996	**53,956**	28,676	18,252	5,643	1,179	206	1996	**50,581**	12,386	18,217	11,084	4,731	4,163
1997	**48,423**	25,655	16,339	5,117	1,113	199	1997	**53,999**	13,155	19,714	11,928	4,875	4,327
1998	**45,159**	24,111	15,333	4,570	947	198	1998	**56,821**	14,254	21,136	12,431	4,918	4,082
1999	**42,765**	23,303	14,293	4,104	882	183	1999	**58,932**	15,571	22,133	12,439	4,939	3,850
2000	**39,861**	21,422	13,596	3,880	799	164	2000	**61,758**	16,352	23,355	13,121	4,959	3,971
2001	**40,370**	22,038	13,597	3,834	732	169	2001	**63,353**	16,747	24,650	13,120	4,931	3,905
2002	**40,385**	22,297	13,394	3,800	727	167	2002	**66,161**	18,088	25,842	13,170	5,009	4,052
2003	**40,569**	22,064	13,811	3,780	751	163	2003	**70,886**	19,811	27,569	14,185	5,096	4,225
2004	**40,955**	22,395	13,716	3,932	777	135	2004	**75,020**	21,196	29,141	14,655	5,617	4,411
2005	**39,721**	22,007	13,099	3,709	759	147	2005	**77,517**	22,284	29,890	15,059	5,751	4,533

1 See section 2.9.

Table 4.1 Live births within marriage: number of previous live-born children and age of mother (five-year age groups), 1995-2005
c. remarried women

England and Wales

Year	Number of previous live-born children[1]						Year	Number of previous live-born children[1]					
	Total	0	1	2	3	4 or more		Total	0	1	2	3	4 or more
	All ages of mother at birth							**25-29**					
1995	**33,285**	9,155	10,836	7,397	3,615	2,282	1995	**7,211**	2,402	2,385	1,493	658	273
1996	**32,582**	9,026	10,557	7,197	3,492	2,310	1996	**6,395**	2,103	2,149	1,318	561	264
1997	**31,445**	8,630	10,249	6,795	3,492	2,279	1997	**5,828**	1,938	1,920	1,155	570	245
1998	**30,185**	8,501	9,983	6,594	3,035	2,072	1998	**5,105**	1,719	1,760	1,018	408	200
1999	**27,489**	7,923	9,056	5,731	2,872	1,907	1999	**4,254**	1,485	1,388	845	382	154
2000	**25,782**	7,299	8,574	5,469	2,591	1,849	2000	**3,719**	1,247	1,228	750	320	174
2001	**23,941**	6,933	8,066	4,966	2,305	1,671	2001	**3,135**	1,083	1,028	604	268	152
2002	**22,825**	6,751	7,717	4,547	2,237	1,573	2002	**2,682**	898	908	508	242	126
2003	**22,599**	6,708	7,597	4,656	2,163	1,475	2003	**2,402**	870	802	452	186	92
2004	**21,534**	6,641	7,169	4,228	2,060	1,436	2004	**2,205**	828	721	379	178	99
2005	**20,042**	6,101	6,534	4,015	1,916	1,476	2005	**2,112**	805	647	388	192	80
	Under 20							**30-34**					
1995	**14**	8	4	2	-	-	1995	**14,030**	4,016	4,748	3,003	1,412	851
1996	**10**	5	3	2	-	-	1996	**13,925**	4,103	4,657	2,959	1,387	819
1997	**7**	5	2	-	-	-	1997	**13,068**	3,798	4,432	2,769	1,313	756
1998	**5**	2	2	-	1	-	1998	**12,389**	3,756	4,189	2,644	1,106	694
1999	**4**	3	1	-	-	-	1999	**11,324**	3,592	3,918	2,202	1,007	605
2000	**5**	4	1	-	-	-	2000	**10,354**	3,196	3,583	2,130	872	573
2001	**2**	1	1	-	-	-	2001	**9,460**	3,051	3,257	1,865	787	500
2002	**8**	4	2	2	-	-	2002	**8,849**	2,947	3,099	1,618	742	443
2003	**7**	3	3	1	-	-	2003	**8,435**	2,808	2,913	1,610	676	428
2004	**3**	-	2	1	-	-	2004	**7,744**	2,613	2,684	1,379	667	401
2005	**5**	3	2	-	-	-	2005	**6,845**	2,302	2,320	1,261	568	394
	20-24							**35 and over**					
1995	**839**	284	303	175	58	19	1995	**11,191**	2,445	3,396	2,724	1,487	1,139
1996	**695**	249	229	146	52	19	1996	**11,557**	2,566	3,519	2,772	1,492	1,208
1997	**645**	236	219	140	40	10	1997	**11,897**	2,653	3,676	2,731	1,569	1,268
1998	**565**	192	204	109	46	14	1998	**12,121**	2,832	3,828	2,823	1,474	1,164
1999	**425**	171	150	71	28	5	1999	**11,482**	2,672	3,599	2,613	1,455	1,143
2000	**401**	149	148	70	28	6	2000	**11,303**	2,703	3,614	2,519	1,371	1,096
2001	**366**	150	126	63	19	8	2001	**10,978**	2,648	3,654	2,434	1,231	1,011
2002	**327**	135	112	66	12	2	2002	**10,959**	2,767	3,596	2,353	1,241	1,002
2003	**318**	123	104	54	26	11	2003	**11,437**	2,904	3,775	2,539	1,275	944
2004	**330**	156	90	62	14	8	2004	**11,252**	3,044	3,672	2,407	1,201	928
2005	**289**	124	104	48	11	2	2005	**10,791**	2,867	3,461	2,318	1,145	1,000

1 See section 2.9.

Table 4.2 Live births within marriage: number of previous
live-born children and age of mother (single years), 2005
a. all married women

England and Wales

Age of mother at birth	Number of previous live-born children[1]											
	Total	0	1	2	3	4	5	6	7	8	9	10 and[2] over
All ages	369,330	155,953	132,007	52,198	18,097	6,383	2,631	1,050	510	248	114	139
Under 16	15	12	2	1	-	-	-	-	-	-	-	-
16	66	59	6	1	-	-	-	-	-	-	-	-
17	310	277	31	2	-	-	-	-	-	-	-	-
18	946	800	127	17	2	-	-	-	-	-	-	-
19	2,317	1,811	461	42	3	-	-	-	-	-	-	-
Under 20	3,654	2,959	627	63	5	-	-	-	-	-	-	-
20	3,947	2,818	970	142	16	1	-	-	-	-	-	-
21	5,678	3,478	1,838	309	48	3	1	-	-	-	-	1
22	7,695	4,266	2,663	635	110	19	2	-	-	-	-	-
23	10,042	5,207	3,480	1,100	221	27	6	1	-	-	-	-
24	12,648	6,362	4,252	1,571	375	75	11	-	-	-	1	1
20-24	40,010	22,131	13,203	3,757	770	125	20	1	-	-	1	2
25	15,555	7,988	4,914	1,941	586	105	18	1	1	1	-	-
26	17,847	9,072	5,656	2,248	651	164	45	6	4	-	-	1
27	19,610	9,865	6,198	2,418	818	226	61	19	3	2	-	-
28	22,051	10,969	7,007	2,768	908	286	85	21	5	1	-	1
29	24,894	12,084	8,288	2,944	1,075	342	115	35	9	2	-	-
25-29	99,957	49,978	32,063	12,319	4,038	1,123	324	82	22	6	-	2
30	27,025	12,702	9,331	3,253	1,122	397	148	48	17	5	2	-
31	28,743	12,690	10,668	3,458	1,243	430	161	59	24	6	1	3
32	28,332	11,472	11,028	3,699	1,349	503	180	67	22	9	1	2
33	27,589	10,126	11,167	4,080	1,353	517	209	89	23	16	7	2
34	25,712	8,744	10,569	4,192	1,321	517	230	76	37	13	9	4
30-34	137,401	55,734	52,763	18,682	6,388	2,364	928	339	123	49	20	11
35	21,809	6,921	8,717	3,889	1,316	539	220	97	61	33	8	8
36	17,980	5,316	7,091	3,525	1,220	457	193	90	42	28	10	8
37	14,460	4,036	5,540	2,945	1,119	418	227	85	45	25	6	14
38	11,080	2,918	4,120	2,330	975	365	187	84	44	24	17	16
39	8,481	2,194	3,102	1,779	778	301	148	81	50	18	13	17
35-39	73,810	21,385	28,570	14,468	5,408	2,080	975	437	242	128	54	63
40	5,856	1,534	2,047	1,195	547	245	130	70	37	23	12	16
41	3,726	968	1,257	756	371	167	97	51	26	14	9	10
42	2,212	519	758	439	238	115	69	30	22	4	7	11
43	1,286	336	356	266	160	73	35	15	16	12	7	10
44	680	174	195	124	91	36	29	8	8	4	1	10
40-44	13,760	3,531	4,613	2,780	1,407	636	360	174	109	57	36	57
45	378	109	90	69	48	28	13	5	7	4	2	3
46	147	46	32	24	18	11	5	4	5	2	-	-
47	85	21	21	18	11	7	3	2	1	-	1	-
48	50	29	10	4	-	1	1	3	-	2	-	-
49 and over	78	30	15	14	4	8	2	3	1	-	-	1
45 and over	738	235	168	129	81	55	24	17	14	8	3	4

Note: 4,155 cases in which age of mother at birth, and 422 cases in which the number of previous live-born children was not stated have been included
with the stated cases (for method of distribution - see sections 3.2 and 3.3).
1 See section 2.9.
2 Detailed distribution for these groups is as follows:

Number of previous live-born children	Number of live births
10	75
11	34
12	16
13	7
14	4
15	2
16	1

Table 4.2 Live births within marriage: number of previous live-born children and age of mother (single years), 2005
b. women married once only

England and Wales

Age of mother at birth	Number of previous live-born children[1]											
	Total	0	1	2	3	4	5	6	7	8	9	10 and[2] over
All ages	349,288	149,852	125,473	48,183	16,181	5,543	2,258	913	441	220	99	125
Under 16	15	12	2	1	-	-	-	-	-	-	-	-
16	66	59	6	1	-	-	-	-	-	-	-	-
17	310	277	31	2	-	-	-	-	-	-	-	-
18	945	799	127	17	2	-	-	-	-	-	-	-
19	2,313	1,809	459	42	3	-	-	-	-	-	-	-
Under 20	3,649	2,956	625	63	5	-	-	-	-	-	-	-
20	3,942	2,816	968	142	15	1	-	-	-	-	-	-
21	5,656	3,467	1,830	306	48	3	1	-	-	-	-	1
22	7,651	4,244	2,653	624	109	19	2	-	-	-	-	-
23	9,954	5,167	3,452	1,087	215	27	5	1	-	-	-	-
24	12,518	6,313	4,196	1,550	372	74	11	-	-	-	1	1
20-24	39,721	22,007	13,099	3,709	759	124	19	1	-	-	1	2
25	15,353	7,909	4,852	1,904	566	102	17	1	1	1	-	-
26	17,547	8,960	5,554	2,192	634	153	44	5	4	-	-	1
27	19,222	9,718	6,079	2,344	785	215	59	18	3	1	-	-
28	21,549	10,775	6,853	2,675	862	275	81	21	5	1	-	1
29	24,174	11,811	8,078	2,816	999	324	103	32	9	2	-	-
25-29	97,845	49,173	31,416	11,931	3,846	1,069	304	77	22	5	-	2
30	26,142	12,397	9,046	3,083	1,043	364	140	46	16	5	2	-
31	27,621	12,286	10,305	3,243	1,163	390	148	53	23	6	1	3
32	26,973	10,967	10,578	3,472	1,248	453	163	62	19	8	1	2
33	25,967	9,582	10,600	3,802	1,201	473	185	81	22	14	6	1
34	23,853	8,200	9,914	3,821	1,165	442	196	63	29	12	7	4
30-34	130,556	53,432	50,443	17,421	5,820	2,122	832	305	109	45	17	10
35	19,988	6,352	8,111	3,523	1,151	475	186	88	56	30	8	8
36	16,157	4,792	6,465	3,145	1,054	381	161	80	38	25	8	8
37	12,787	3,573	4,963	2,589	968	344	199	74	40	24	3	10
38	9,600	2,530	3,636	2,014	805	301	151	72	40	19	16	16
39	7,222	1,896	2,702	1,500	636	224	114	71	36	16	13	14
35-39	65,754	19,143	25,877	12,771	4,614	1,725	811	385	210	114	48	56
40	4,864	1,293	1,751	968	435	188	101	54	27	22	11	14
41	3,057	817	1,067	606	290	113	79	37	22	10	8	8
42	1,753	423	621	333	173	91	52	23	16	4	6	11
43	1,010	282	293	197	116	50	21	13	15	10	4	9
44	511	129	155	90	69	22	19	5	8	4	1	9
40-44	11,195	2,944	3,887	2,194	1,083	464	272	132	88	50	30	51
45	289	92	67	55	29	19	12	2	5	3	2	3
46	106	35	22	15	13	9	3	3	5	1	-	-
47	69	19	17	13	8	6	2	2	1	-	1	-
48	41	22	8	4	-	1	1	3	-	2	-	-
49 and over	63	29	12	7	4	4	2	3	1	-	-	1
45 and over	568	197	126	94	54	39	20	13	12	6	3	4

Note: 4,146 cases in which age of mother at birth, and 422 cases in which the number of previous live-born children was not stated have been included with the stated cases (for method of distribution - see sections 3.2 and 3.3).
1 See section 2.9.
2 Detailed distribution for these groups is as follows:

Number of previous live-born children	Number of live births
10	68
11	30
12	14
13	6
14	4
15	2
16	1

Table 4.2   Live births within marriage: number of previous
live-born children and age of mother (single years), 2005
c. remarried women

England and Wales

Age of mother at birth	Number of previous live-born children[1]											
	Total	0	1	2	3	4	5	6	7	8	9	10 and[2] over
All ages	20,042	6,101	6,534	4,015	1,916	840	373	137	69	28	15	14
Under 16	-	-	-	-	-	-	-	-	-	-	-	-
16	-	-	-	-	-	-	-	-	-	-	-	-
17	-	-	-	-	-	-	-	-	-	-	-	-
18	1	1	-	-	-	-	-	-	-	-	-	-
19	4	2	2	-	-	-	-	-	-	-	-	-
Under 20	5	3	2	-	-	-	-	-	-	-	-	-
20	5	2	2	-	1	-	-	-	-	-	-	-
21	22	11	8	3	-	-	-	-	-	-	-	-
22	44	22	10	11	1	-	-	-	-	-	-	-
23	88	40	28	13	6	-	1	-	-	-	-	-
24	130	49	56	21	3	1	-	-	-	-	-	-
20-24	289	124	104	48	11	1	1	-	-	-	-	-
25	202	79	62	37	20	3	1	-	-	-	-	-
26	300	112	102	56	17	11	1	1	-	-	-	-
27	388	147	119	74	33	11	2	1	-	1	-	-
28	502	194	154	93	46	11	4	-	-	-	-	-
29	720	273	210	128	76	18	12	3	-	-	-	-
25-29	2,112	805	647	388	192	54	20	5	-	1	-	-
30	883	305	285	170	79	33	8	2	1	-	-	-
31	1,122	404	363	215	80	40	13	6	1	-	-	-
32	1,359	505	450	227	101	50	17	5	3	1	-	-
33	1,622	544	567	278	152	44	24	8	1	2	1	1
34	1,859	544	655	371	156	75	34	13	8	1	2	-
30-34	6,845	2,302	2,320	1,261	568	242	96	34	14	4	3	1
35	1,821	569	606	366	165	64	34	9	5	3	-	-
36	1,823	524	626	380	166	76	32	10	4	3	2	-
37	1,673	463	577	356	151	74	28	11	5	1	3	4
38	1,480	388	484	316	170	64	36	12	4	5	1	-
39	1,259	298	400	279	142	77	34	10	14	2	-	3
35-39	8,056	2,242	2,693	1,697	794	355	164	52	32	14	6	7
40	992	241	296	227	112	57	29	16	10	1	1	2
41	669	151	190	150	81	54	18	14	4	4	1	2
42	459	96	137	106	65	24	17	7	6	-	1	-
43	276	54	63	69	44	23	14	2	1	2	3	1
44	169	45	40	34	22	14	10	3	-	-	-	1
40-44	2,565	587	726	586	324	172	88	42	21	7	6	6
45	89	17	23	14	19	9	1	3	2	1	-	-
46	41	11	10	9	5	2	2	1	-	1	-	-
47	16	2	4	5	3	1	1	-	-	-	-	-
48	9	7	2	-	-	-	-	-	-	-	-	-
49 and over	15	1	3	7	-	4	-	-	-	-	-	-
45 and over	170	38	42	35	27	16	4	4	2	2	-	-

Note: 9 cases in which age of mother at birth was not stated have been included with the stated cases (for method of distribution - see sections 3.2 and 3.3).
1 See section 2.9.
2 Detailed distribution for these groups is as follows:

Number of previous live-born children	Number of live births
10	7
11	4
12	2
13	1

Table 4.3   Live births and stillbirths within marriage (numbers and sex ratios):   **England and Wales**
number of previous live-born children, age of mother and sex, 2005

Age of mother at birth	Sex	Number of previous live-born children[1]											
		Total	0	1	2	3	4	5	6	7	8	9	10 and[2] over
		Numbers											
All ages	M	190,341	80,513	67,904	26,892	9,301	3,244	1,408	547	262	126	64	80
	F	180,850	76,281	64,692	25,551	8,905	3,171	1,246	515	251	124	51	63
Under 20	M	1,926	1,584	313	26	3	-	-	-	-	-	-	-
	F	1,754	1,398	317	37	2	-	-	-	-	-	-	-
20-24	M	20,610	11,344	6,870	1,935	390	56	13	1	-	-	-	1
	F	19,623	10,910	6,411	1,836	388	69	7	-	-	-	1	1
25-29	M	51,391	25,838	16,428	6,291	2,062	564	161	32	11	2	-	2
	F	49,077	24,388	15,803	6,097	1,993	562	166	52	12	4	-	-
30-34	M	70,769	28,779	26,994	9,661	3,321	1,213	529	180	55	22	9	6
	F	67,241	27,237	25,951	9,096	3,111	1,166	406	163	68	27	11	5
35-39	M	38,174	11,019	14,829	7,429	2,805	1,069	518	241	132	67	31	34
	F	36,020	10,492	13,874	7,105	2,634	1,023	467	199	112	61	23	30
40-44	M	7,080	1,813	2,379	1,483	685	318	175	84	55	32	22	34
	F	6,780	1,754	2,258	1,316	731	320	188	93	54	27	15	24
45 and over	M	391	136	91	67	35	24	12	9	9	3	2	3
	F	355	102	78	64	46	31	12	8	5	5	1	3
Ratio: male births inside marriage per 1,000 female births inside marriage													
All ages		1,052	1,055	1,050	1,052	1,044	1,023	1,130	1,062	1,044	1,016	1,255	1,300
Under 20		1,098	1,133	987	703	1,500	-	-	-	-	-	-	-
20-24		1,050	1,040	1,072	1,054	1,005	812	1,857	-	-	-	-	1,000
25-29		1,047	1,059	1,040	1,032	1,035	1,004	970	615	917	500	-	-
30-34		1,052	1,057	1,040	1,062	1,068	1,040	1,303	1,104	809	815	818	1,200
35-39		1,060	1,050	1,069	1,046	1,065	1,045	1,109	1,211	1,179	1,098	1,348	1,133
40-44		1,044	1,034	1,054	1,127	937	994	931	903	1,019	1,185	1,467	1,455
45 and over		1,101	1,333	1,167	1,047	761	774	1,000	1,125	1,800	600	2,000	1,500

Note: 4,373 cases in which age of mother at birth, and 702 cases in which the number of previous live-born children was not stated have been included with the stated cases (for method of distribution - see sections 3.2 and 3.3).
1 See section 2.9.
2 Detailed distribution for these groups is as follows:

Number of previous live-born children	Number of births
10	76
11	35
12	16
13	7
14	4
15	4
16	1

Table 5.1 First live births within marriage: duration
of marriage and age of mother, 1995-2005
a. all married women

England and Wales

Year	All durations	Completed months		Completed years													
		0-7	8-11	0	1	2	3	4	5	6	7	8	9	10-14	15-19	20 and over
All ages of mother at birth																
1995	**168,118**	17,332	15,611	32,943	40,542	28,659	19,683	13,947	10,115	6,788	4,763	3,098	2,262	4,411	797	110
1996	**163,020**	17,300	15,489	32,789	39,521	27,283	18,951	13,225	9,333	6,806	4,622	3,129	2,173	4,274	807	107
1997	**157,049**	16,760	15,199	31,959	38,650	26,225	18,075	12,479	8,680	6,243	4,632	3,043	2,181	4,044	728	110
1998	**155,708**	16,582	15,617	32,199	38,631	26,169	17,648	12,097	8,643	5,886	4,350	3,091	2,006	4,165	720	103
1999	**153,423**	16,193	15,346	31,539	38,723	25,803	17,267	11,989	8,271	5,797	4,004	2,964	2,223	4,085	670	88
2000	**146,509**	14,423	14,838	29,261	38,069	25,331	16,239	11,166	7,561	5,422	3,858	2,734	2,046	4,165	575	82
2001	**143,908**	13,565	14,887	28,452	37,897	25,564	16,428	10,706	7,179	4,961	3,653	2,571	1,866	3,959	585	87
2002	**145,241**	13,871	14,904	28,775	38,634	26,372	16,566	10,889	7,265	4,961	3,451	2,396	1,736	3,544	539	113
2003	**150,982**	14,489	15,990	30,479	39,795	26,841	17,546	11,518	7,579	4,995	3,516	2,420	1,783	3,791	621	98
2004	**154,548**	14,545	16,998	31,543	41,891	26,926	17,997	11,670	7,583	5,155	3,415	2,385	1,646	3,661	584	92
2005	**155,953**	13,624	17,097	30,721	43,609	27,895	17,359	11,616	7,935	5,180	3,402	2,323	1,644	3,534	635	100
Under 20																
1995	**4,308**	1,605	1,128	2,733	1,228	282	56	7	-	1	1	-	-	-	-	-
1996	**4,223**	1,567	1,149	2,716	1,180	256	61	8	2	-	-	-	-	-	-	-
1997	**4,121**	1,505	1,031	2,536	1,237	272	71	4	1	-	-	-	-	-	-	-
1998	**4,188**	1,490	1,123	2,613	1,223	280	63	7	2	-	-	-	-	-	-	-
1999	**4,290**	1,455	996	2,451	1,430	333	69	6	1	-	-	-	-	-	-	-
2000	**3,810**	1,225	903	2,128	1,221	383	75	3	-	-	-	-	-	-	-	-
2001	**3,781**	1,086	998	2,084	1,253	358	78	6	2	-	-	-	-	-	-	-
2002	**3,817**	1,119	888	2,007	1,366	363	73	8	-	-	-	-	-	-	-	-
2003	**3,471**	1,033	902	1,935	1,149	307	72	7	-	1	-	-	-	-	-	-
2004	**3,333**	1,010	844	1,854	1,123	282	67	5	2	-	-	-	-	-	-	-
2005	**2,959**	868	695	1,563	1,016	298	71	8	2	1	-	-	-	-	-	-
20-24																
1995	**32,340**	5,783	4,755	10,538	10,893	6,153	2,964	1,215	407	122	32	6	4	4	-	-
1996	**28,925**	5,408	4,469	9,877	9,780	5,227	2,510	1,033	347	107	34	4	1	5	-	-
1997	**25,891**	4,913	4,135	9,048	8,733	4,629	2,113	910	305	112	28	8	3	2	-	-
1998	**24,303**	4,494	3,885	8,379	8,270	4,386	1,973	832	299	112	35	8	-	9	-	-
1999	**23,474**	4,336	3,706	8,042	8,000	4,165	1,952	803	353	100	36	7	7	9	-	-
2000	**21,571**	3,896	3,415	7,311	7,507	4,016	1,665	681	263	78	28	15	-	7	-	-
2001	**22,188**	3,755	3,514	7,269	7,756	4,233	1,817	696	272	100	34	6	1	4	-	-
2002	**22,432**	3,620	3,566	7,186	7,987	4,236	1,819	726	292	143	32	8	2	1	-	-
2003	**22,187**	3,815	3,613	7,428	7,594	4,091	1,843	761	303	120	29	12	-	6	-	-
2004	**22,551**	3,865	3,676	7,541	7,778	3,989	1,913	817	335	115	41	9	5	8	-	-
2005	**22,131**	3,477	3,618	7,095	7,805	4,072	1,847	772	361	119	47	6	2	5	-	-
25-29																
1995	**71,036**	5,391	5,850	11,241	17,299	13,629	10,082	7,518	5,234	3,071	1,709	758	309	186	-	-
1996	**67,178**	5,396	5,695	11,091	16,590	12,889	9,552	6,846	4,604	2,915	1,530	699	286	176	-	-
1997	**63,139**	5,280	5,620	10,900	16,230	12,182	8,905	6,052	4,055	2,405	1,397	649	236	127	1	-
1998	**60,602**	5,377	5,804	11,181	15,895	11,552	8,269	5,593	3,746	2,137	1,238	609	230	148	4	-
1999	**57,385**	5,152	5,611	10,763	15,467	11,093	7,596	5,229	3,298	1,961	1,073	561	226	117	1	-
2000	**52,673**	4,352	5,306	9,658	14,859	10,595	7,010	4,538	2,740	1,606	871	427	220	149	-	-
2001	**48,849**	4,092	5,084	9,176	13,945	10,028	6,437	3,997	2,401	1,384	778	375	188	136	4	-
2002	**47,111**	4,060	4,967	9,027	13,592	9,802	6,310	3,763	2,177	1,208	630	324	157	121	-	-
2003	**48,439**	4,079	5,212	9,291	14,149	9,935	6,567	3,856	2,239	1,145	635	318	168	134	2	-
2004	**48,890**	4,190	5,656	9,846	14,634	9,798	6,506	3,729	2,119	1,156	536	298	136	130	2	-
2005	**49,978**	4,024	5,823	9,847	15,470	10,122	6,256	3,835	2,164	1,174	562	296	123	127	2	-
30 and over																
1995	**60,434**	4,553	3,878	8,431	11,122	8,595	6,581	5,207	4,474	3,594	3,021	2,334	1,947	4,221	797	110
1996	**62,694**	4,929	4,176	9,105	11,971	8,911	6,828	5,338	4,380	3,784	3,058	2,426	1,886	4,093	807	107
1997	**63,898**	5,062	4,413	9,475	12,450	9,142	6,986	5,513	4,319	3,726	3,207	2,386	1,942	3,915	727	110
1998	**66,615**	5,221	4,805	10,026	13,243	9,951	7,343	5,665	4,596	3,637	3,077	2,474	1,776	4,008	716	103
1999	**68,274**	5,250	5,033	10,283	13,826	10,212	7,650	5,951	4,619	3,736	2,895	2,396	1,990	3,959	669	88
2000	**68,455**	4,950	5,214	10,164	14,482	10,337	7,489	5,944	4,558	3,738	2,959	2,292	1,826	4,009	575	82
2001	**69,090**	4,632	5,291	9,923	14,943	10,945	8,096	6,007	4,504	3,477	2,841	2,190	1,677	3,819	581	87
2002	**71,881**	5,072	5,483	10,555	15,689	11,971	8,364	6,392	4,796	3,610	2,789	2,064	1,577	3,422	539	113
2003	**76,885**	5,562	6,263	11,825	16,903	12,508	9,064	6,894	5,037	3,729	2,852	2,090	1,615	3,651	619	98
2004	**79,774**	5,480	6,822	12,302	18,356	12,857	9,511	7,119	5,127	3,884	2,838	2,078	1,505	3,523	582	92
2005	**80,885**	5,255	6,961	12,216	19,318	13,403	9,185	7,001	5,408	3,886	2,793	2,021	1,519	3,402	633	100

Table 5.1   First live births within marriage: duration
of marriage and age of mother, 1995-2005
b. women married once only

England and Wales

Year	All durations	Completed months		Completed years										10-14	15-19	20 and over
		0-7	8-11	0	1	2	3	4	5	6	7	8	9			
All ages of mother at birth																
1995	158,963	15,368	14,320	29,688	38,073	27,298	18,947	13,461	9,799	6,609	4,660	3,023	2,200	4,304	791	110
1996	153,994	15,351	14,225	29,576	36,982	26,052	18,207	12,779	9,040	6,616	4,491	3,059	2,124	4,166	797	105
1997	148,419	14,878	14,049	28,927	36,178	24,992	17,418	12,048	8,401	6,069	4,526	2,966	2,128	3,938	719	109
1998	147,207	14,745	14,365	29,110	36,324	24,909	16,981	11,691	8,390	5,715	4,223	3,023	1,964	4,059	716	102
1999	145,500	14,545	14,186	28,731	36,489	24,678	16,594	11,595	8,010	5,653	3,914	2,895	2,181	4,007	665	88
2000	139,210	12,980	13,791	26,771	35,999	24,278	15,601	10,786	7,318	5,268	3,766	2,655	2,012	4,105	570	81
2001	136,975	12,323	13,894	26,217	35,888	24,468	15,807	10,348	6,975	4,823	3,567	2,510	1,829	3,876	580	87
2002	138,490	12,540	13,993	26,533	36,704	25,349	15,992	10,521	7,042	4,826	3,356	2,333	1,696	3,494	534	110
2003	144,274	13,166	15,025	28,191	37,893	25,844	16,951	11,190	7,351	4,868	3,428	2,362	1,753	3,731	616	96
2004	147,907	13,328	15,987	29,315	39,984	25,954	17,422	11,300	7,391	5,013	3,328	2,332	1,599	3,602	576	91
2005	149,852	12,512	16,160	28,672	41,831	26,958	16,886	11,315	7,742	5,049	3,343	2,264	1,607	3,461	624	100
Under 20																
1995	4,300	1,600	1,128	2,728	1,225	282	56	7	-	1	1	-	-	-	-	-
1996	4,218	1,565	1,146	2,711	1,180	256	61	8	2	-	-	-	-	-	-	-
1997	4,116	1,502	1,029	2,531	1,237	272	71	4	1	-	-	-	-	-	-	-
1998	4,186	1,489	1,123	2,612	1,222	280	63	7	2	-	-	-	-	-	-	-
1999	4,287	1,454	994	2,448	1,430	333	69	6	1	-	-	-	-	-	-	-
2000	3,806	1,223	902	2,125	1,220	383	75	3	-	-	-	-	-	-	-	-
2001	3,780	1,086	997	2,083	1,253	358	78	6	2	-	-	-	-	-	-	-
2002	3,813	1,118	888	2,006	1,364	362	73	8	-	-	-	-	-	-	-	-
2003	3,468	1,031	901	1,932	1,149	307	72	7	-	1	-	-	-	-	-	-
2004	3,333	1,010	844	1,854	1,123	282	67	5	2	-	-	-	-	-	-	-
2005	2,956	867	695	1,562	1,014	298	71	8	2	1	-	-	-	-	-	-
20-24																
1995	32,056	5,672	4,682	10,354	10,820	6,134	2,958	1,214	406	122	32	6	6	4	-	-
1996	28,676	5,299	4,421	9,720	9,713	5,210	2,502	1,033	347	107	34	4	1	5	-	-
1997	25,655	4,824	4,085	8,909	8,662	4,613	2,107	907	304	112	28	8	3	2	-	-
1998	24,111	4,423	3,836	8,259	8,221	4,368	1,969	831	299	112	35	8	-	9	-	-
1999	23,303	4,269	3,660	7,929	7,956	4,158	1,946	802	353	100	36	7	7	9	-	-
2000	21,422	3,848	3,375	7,223	7,461	4,006	1,661	680	263	78	28	15	-	7	-	-
2001	22,038	3,704	3,474	7,178	7,720	4,212	1,816	695	272	100	34	6	1	4	-	-
2002	22,297	3,572	3,535	7,107	7,948	4,224	1,815	725	292	143	32	8	2	1	-	-
2003	22,064	3,784	3,578	7,362	7,553	4,081	1,837	761	303	120	29	12	-	6	-	-
2004	22,395	3,813	3,632	7,445	7,735	3,981	1,908	814	335	114	41	9	5	8	-	-
2005	22,007	3,434	3,586	7,020	7,766	4,065	1,844	772	361	119	47	6	2	5	-	-
25-29																
1995	68,634	4,744	5,414	10,158	16,525	13,309	9,970	7,447	5,208	3,060	1,707	757	308	185	-	-
1996	65,075	4,817	5,310	10,127	15,934	12,625	9,427	6,791	4,585	2,905	1,526	699	285	171	-	-
1997	61,201	4,754	5,280	10,034	15,572	11,940	8,809	6,012	4,036	2,395	1,393	648	236	125	1	-
1998	58,883	4,890	5,458	10,348	15,386	11,338	8,169	5,557	3,734	2,130	1,233	609	230	145	4	-
1999	55,900	4,734	5,289	10,023	15,027	10,914	7,530	5,196	3,278	1,956	1,071	561	226	117	1	-
2000	51,426	4,035	5,067	9,102	14,457	10,428	6,945	4,502	2,729	1,601	869	426	220	147	-	-
2001	47,766	3,824	4,866	8,690	13,589	9,892	6,373	3,974	2,391	1,379	775	375	188	136	4	-
2002	46,213	3,801	4,811	8,612	13,305	9,696	6,261	3,737	2,168	1,206	628	322	157	121	-	-
2003	47,569	3,838	5,033	8,871	13,905	9,821	6,512	3,831	2,232	1,143	632	318	168	134	2	-
2004	48,062	3,994	5,469	9,463	14,355	9,707	6,462	3,705	2,114	1,155	536	298	136	129	2	-
2005	49,173	3,824	5,644	9,468	15,210	10,011	6,225	3,820	2,157	1,173	562	295	123	127	2	-
30 and over																
1995	53,973	3,352	3,096	6,448	9,503	7,573	5,963	4,793	4,185	3,426	2,920	2,260	1,886	4,115	791	110
1996	56,025	3,670	3,348	7,018	10,155	7,961	6,217	4,947	4,106	3,604	2,931	2,356	1,838	3,990	797	105
1997	57,447	3,798	3,655	7,453	10,707	8,167	6,431	5,125	4,060	3,562	3,105	2,310	1,889	3,811	718	109
1998	60,027	3,943	3,948	7,891	11,495	8,923	6,780	5,296	4,355	3,473	2,955	2,406	1,734	3,905	712	102
1999	62,010	4,088	4,243	8,331	12,076	9,273	7,049	5,591	4,378	3,597	2,807	2,327	1,948	3,881	664	88
2000	62,556	3,874	4,447	8,321	12,861	9,461	6,920	5,601	4,326	3,589	2,869	2,214	1,792	3,951	570	81
2001	63,391	3,709	4,557	8,266	13,326	10,006	7,540	5,673	4,310	3,344	2,758	2,129	1,640	3,736	576	87
2002	66,167	4,049	4,759	8,808	14,087	11,067	7,843	6,051	4,582	3,477	2,696	2,003	1,537	3,372	534	110
2003	71,173	4,513	5,513	10,026	15,286	11,635	8,530	6,591	4,816	3,604	2,767	2,032	1,585	3,591	614	96
2004	74,117	4,511	6,042	10,553	16,771	11,984	8,985	6,776	4,940	3,744	2,751	2,025	1,458	3,465	574	91
2005	75,716	4,387	6,235	10,622	17,841	12,584	8,746	6,715	5,222	3,756	2,734	1,963	1,482	3,329	622	100

Table 5.1　First live births within marriage: duration
of marriage and age of mother, 1995-2005
c. remarried women

England and Wales

Year	Duration of current marriage																
	All durations	Completed months		Completed years													
		0-7	8-11	0	1	2	3	4	5	6	7	8	9	10-14	15-19	20 and over
All ages of mother at birth																
1995	**9,155**	1,964	1,291	3,255	2,469	1,361	736	486	316	179	103	75	62	107	6	-
1996	**9,026**	1,949	1,264	3,213	2,539	1,231	744	446	293	190	131	70	49	108	10	2
1997	**8,630**	1,882	1,150	3,032	2,472	1,233	657	431	279	174	106	77	53	106	9	1
1998	**8,501**	1,837	1,252	3,089	2,307	1,260	667	406	253	171	127	68	42	106	4	1
1999	**7,923**	1,648	1,160	2,808	2,234	1,125	673	394	261	144	90	69	42	78	5	-
2000	**7,299**	1,443	1,047	2,490	2,070	1,053	638	380	243	154	92	79	34	60	5	1
2001	**6,933**	1,242	993	2,235	2,009	1,096	621	358	204	138	86	61	37	83	5	-
2002	**6,751**	1,331	911	2,242	1,930	1,023	574	368	223	135	95	63	40	50	5	3
2003	**6,708**	1,323	965	2,288	1,902	997	595	328	228	127	88	58	30	60	5	2
2004	**6,641**	1,217	1,011	2,228	1,907	972	575	370	192	142	87	53	47	59	8	1
2005	**6,101**	1,112	937	2,049	1,778	937	473	301	193	131	59	59	37	73	11	-
Under 20																
1995	**8**	5	-	5	3	-	-	-	-	-	-	-	-	-	-	-
1996	**5**	2	3	5	-	-	-	-	-	-	-	-	-	-	-	-
1997	**5**	3	2	5	-	-	-	-	-	-	-	-	-	-	-	-
1998	**2**	1	-	1	1	-	-	-	-	-	-	-	-	-	-	-
1999	**3**	1	2	3	-	-	-	-	-	-	-	-	-	-	-	-
2000	**4**	2	1	3	1	-	-	-	-	-	-	-	-	-	-	-
2001	**1**	-	1	1	-	-	-	-	-	-	-	-	-	-	-	-
2002	**4**	1	-	1	2	1	-	-	-	-	-	-	-	-	-	-
2003	**3**	2	1	3	-	-	-	-	-	-	-	-	-	-	-	-
2004	**-**	-	-	-	-	-	-	-	-	-	-	-	-	-	-	-
2005	**3**	1	-	1	2	-	-	-	-	-	-	-	-	-	-	-
20-24																
1995	**284**	111	73	184	73	19	6	1	1	-	-	-	-	-	-	-
1996	**249**	109	48	157	67	17	8	-	-	-	-	-	-	-	-	-
1997	**236**	89	50	139	71	16	6	3	1	-	-	-	-	-	-	-
1998	**192**	71	49	120	49	18	4	1	-	-	-	-	-	-	-	-
1999	**171**	67	46	113	44	7	6	1	-	-	-	-	-	-	-	-
2000	**149**	48	40	88	46	10	4	1	-	-	-	-	-	-	-	-
2001	**150**	51	40	91	36	21	1	1	-	-	-	-	-	-	-	-
2002	**135**	48	31	79	39	12	4	1	-	-	-	-	-	-	-	-
2003	**123**	31	35	66	41	10	6	-	-	-	-	-	-	-	-	-
2004	**156**	52	44	96	43	8	5	3	-	1	-	-	-	-	-	-
2005	**124**	43	32	75	39	7	3	-	-	-	-	-	-	-	-	-
25-29																
1995	**2,402**	647	436	1,083	774	320	112	71	26	11	2	1	1	1	-	-
1996	**2,103**	579	385	964	656	264	125	55	19	10	4	-	1	5	-	-
1997	**1,938**	526	340	866	658	242	96	40	19	10	4	1	-	2	-	-
1998	**1,719**	487	346	833	509	214	100	36	12	7	5	-	-	3	-	-
1999	**1,485**	418	322	740	440	179	66	33	20	5	2	-	-	-	-	-
2000	**1,247**	317	239	556	402	167	65	36	11	5	2	1	-	2	-	-
2001	**1,083**	268	218	486	356	136	64	23	10	5	3	-	-	-	-	-
2002	**898**	259	156	415	287	106	49	26	9	2	2	2	-	-	-	-
2003	**870**	241	179	420	244	114	55	25	7	2	3	-	-	-	-	-
2004	**828**	196	187	383	279	91	44	24	5	1	-	-	-	1	-	-
2005	**805**	200	179	379	260	111	31	15	7	1	-	1	-	-	-	-
30 and over																
1995	**6,461**	1,201	782	1,983	1,619	1,022	618	414	289	168	101	74	61	106	6	-
1996	**6,669**	1,259	828	2,087	1,816	950	611	391	274	180	127	70	48	103	10	2
1997	**6,451**	1,264	758	2,022	1,743	975	555	388	259	164	102	76	53	104	9	1
1998	**6,588**	1,278	857	2,135	1,748	1,028	563	369	241	164	122	68	42	103	4	1
1999	**6,264**	1,162	790	1,952	1,750	939	601	360	241	139	88	69	42	78	5	-
2000	**5,899**	1,076	767	1,843	1,621	876	569	343	232	149	90	78	34	58	5	1
2001	**5,699**	923	734	1,657	1,617	939	556	334	194	133	83	61	37	83	5	-
2002	**5,714**	1,023	724	1,747	1,602	904	521	341	214	133	93	61	40	50	5	3
2003	**5,712**	1,049	750	1,799	1,617	873	534	303	221	125	85	58	30	60	5	2
2004	**5,657**	969	780	1,749	1,585	873	526	343	187	140	87	53	47	58	8	1
2005	**5,169**	868	726	1,594	1,477	819	439	286	186	130	59	58	37	73	11	-

Table 5.2   Live births within 8 months of marriage: **England and Wales**
order of marriage and age of mother, 1995-2005

Age of mother at birth	1995	1996	1997	1998	1999	2000	2001	2002	2003	2004	2005
All married women											
All ages	**25,454**	**25,265**	**24,467**	**24,042**	**23,196**	**20,806**	**19,679**	**19,745**	**20,561**	**20,791**	**19,294**
Under 16	2	1	4	1	1	-	-	4	1	4	3
16	61	86	73	48	54	42	57	48	37	27	29
17	250	225	221	223	201	156	156	157	134	113	104
18	587	546	595	569	552	467	401	429	376	382	318
19	960	925	818	861	842	737	618	611	639	631	524
Under 20	**1,860**	**1,783**	**1,711**	**1,702**	**1,650**	**1,402**	**1,232**	**1,249**	**1,187**	**1,157**	**978**
20	1,238	1,139	1,010	913	988	904	868	805	763	791	688
21	1,416	1,306	1,167	1,145	1,059	1,033	987	920	1,025	917	825
22	1,590	1,448	1,274	1,202	1,080	975	1,041	1,042	957	1,056	909
23	1,681	1,608	1,408	1,296	1,204	1,043	1,063	1,028	1,179	1,158	1,056
24	1,762	1,684	1,603	1,453	1,353	1,163	1,031	1,024	1,059	1,182	1,122
20-24	**7,687**	**7,185**	**6,462**	**6,009**	**5,684**	**5,118**	**4,990**	**4,819**	**4,983**	**5,104**	**4,600**
25-29	7,979	8,035	7,708	7,612	7,294	6,266	5,838	5,609	5,692	5,800	5,526
30-34	5,370	5,531	5,766	5,656	5,541	5,087	4,824	5,007	5,163	5,233	4,846
35-39	2,121	2,264	2,310	2,546	2,501	2,432	2,270	2,479	2,858	2,814	2,640
40 and over	437	467	510	517	526	501	525	582	678	683	704
Women married once only											
All ages	**20,250**	**20,175**	**19,596**	**19,438**	**19,169**	**17,210**	**16,530**	**16,613**	**17,389**	**17,864**	**16,754**
Under 16	2	1	4	1	1	-	-	3	1	4	3
16	61	86	73	48	54	42	57	48	37	27	29
17	250	225	220	222	201	156	156	157	134	113	104
18	584	546	595	568	552	467	401	429	374	381	317
19	956	919	815	861	841	735	618	610	637	631	524
Under 20	**1,853**	**1,777**	**1,707**	**1,700**	**1,649**	**1,400**	**1,232**	**1,247**	**1,183**	**1,156**	**977**
20	1,232	1,134	1,003	906	980	899	867	798	757	789	687
21	1,395	1,280	1,146	1,130	1,045	1,024	979	906	1,015	905	818
22	1,548	1,406	1,251	1,177	1,061	948	1,021	1,031	945	1,040	897
23	1,582	1,523	1,357	1,252	1,162	1,011	1,030	1,002	1,152	1,126	1,028
24	1,628	1,570	1,500	1,367	1,285	1,099	982	981	1,031	1,138	1,094
20-24	**7,385**	**6,913**	**6,257**	**5,832**	**5,533**	**4,981**	**4,879**	**4,718**	**4,900**	**4,998**	**4,524**
25-29	6,440	6,604	6,461	6,511	6,370	5,516	5,223	5,025	5,168	5,359	5,107
30-34	3,362	3,538	3,796	3,856	3,935	3,634	3,579	3,802	3,970	4,119	3,944
35-39	1,023	1,143	1,158	1,344	1,443	1,434	1,352	1,527	1,801	1,861	1,808
40 and over	187	200	217	195	239	245	265	294	367	371	394
Remarried women											
All ages	**5,204**	**5,090**	**4,871**	**4,604**	**4,027**	**3,596**	**3,149**	**3,132**	**3,172**	**2,927**	**2,540**
Under 16	-	-	-	-	-	-	-	1	-	-	-
16	-	-	-	-	-	-	-	-	-	-	-
17	-	-	1	1	-	-	-	-	-	-	-
18	3	-	-	1	-	-	-	-	2	1	1
19	4	6	3	-	1	2	-	1	2	-	-
Under 20	**7**	**6**	**4**	**2**	**1**	**2**	**-**	**2**	**4**	**1**	**1**
20	6	5	7	7	8	5	1	7	6	2	1
21	21	26	21	15	14	9	8	14	10	12	7
22	42	42	23	25	19	27	20	11	12	16	12
23	99	85	51	44	42	32	33	26	27	32	28
24	134	114	103	86	68	64	49	43	28	44	28
20-24	**302**	**272**	**205**	**177**	**151**	**137**	**111**	**101**	**83**	**106**	**76**
25-29	1,539	1,431	1,247	1,101	924	750	615	584	524	441	419
30-34	2,008	1,993	1,970	1,800	1,606	1,453	1,245	1,205	1,193	1,114	902
35-39	1,098	1,121	1,152	1,202	1,058	998	918	952	1,057	953	832
40 and over	250	267	293	322	287	256	260	288	311	312	310

Table 5.3   Live births within 8 months of marriage: duration of marriage, order of marriage and age of mother, 2005   **England and Wales**

Age of mother at birth	Duration of current marriage - completed months								
	0-7	0	1	2	3	4	5	6	7
All married women									
All ages	**19,294**	**705**	**1,263**	**1,833**	**2,570**	**3,174**	**3,520**	**3,168**	**3,061**
Under 16	**3**	-	-	2	-	-	-	-	1
16	**29**	4	3	6	4	2	7	2	1
17	**104**	4	13	10	18	12	20	13	14
18	**318**	23	36	45	57	46	50	34	27
19	**524**	29	48	54	80	93	72	80	68
Under 20	**978**	**60**	**100**	**117**	**159**	**153**	**149**	**129**	**111**
20	**688**	33	46	82	92	115	133	111	76
21	**825**	37	46	83	106	140	137	129	147
22	**909**	32	72	92	103	160	169	137	144
23	**1,056**	40	76	92	161	156	201	152	178
24	**1,122**	53	80	112	155	176	181	201	164
20-24	**4,600**	**195**	**320**	**461**	**617**	**747**	**821**	**730**	**709**
25-29	**5,526**	188	318	492	664	905	1,039	966	954
30-34	**4,846**	150	283	429	631	784	924	835	810
35-39	**2,640**	78	172	250	392	452	475	422	399
40 and over	**704**	34	70	84	107	133	112	86	78
Women married once only									
All ages	**16,754**	**614**	**1,066**	**1,580**	**2,255**	**2,760**	**3,069**	**2,744**	**2,666**
Under 16	**3**	-	-	2	-	-	-	-	1
16	**29**	4	3	6	4	2	7	2	1
17	**104**	4	13	10	18	12	20	13	14
18	**317**	23	36	45	57	46	49	34	27
19	**524**	29	48	54	80	93	72	80	68
Under 20	**977**	**60**	**100**	**117**	**159**	**153**	**148**	**129**	**111**
20	**687**	33	46	82	92	115	132	111	76
21	**818**	37	44	83	105	138	136	128	147
22	**897**	31	69	89	103	159	167	137	142
23	**1,028**	39	71	87	157	155	197	148	174
24	**1,094**	51	76	107	151	175	176	198	160
20-24	**4,524**	**191**	**306**	**448**	**608**	**742**	**808**	**722**	**699**
25-29	**5,107**	162	292	456	616	841	976	884	880
30-34	**3,944**	119	216	345	514	653	743	675	679
35-39	**1,808**	62	114	172	292	304	321	287	256
40 and over	**394**	20	38	42	66	67	73	47	41
Remarried women									
All ages	**2,540**	**91**	**197**	**253**	**315**	**414**	**451**	**424**	**395**
Under 16	-	-	-	-	-	-	-	-	-
16	-	-	-	-	-	-	-	-	-
17	-	-	-	-	-	-	-	-	-
18	**1**	-	-	-	-	-	1	-	-
19	-	-	-	-	-	-	-	-	-
Under 20	**1**	-	-	-	-	-	**1**	-	-
20	**1**	-	-	-	-	-	1	-	-
21	**7**	-	2	-	1	2	1	1	-
22	**12**	1	3	3	-	1	2	-	2
23	**28**	1	5	5	4	1	4	4	4
24	**28**	2	4	5	4	1	5	3	4
20-24	**76**	**4**	**14**	**13**	**9**	**5**	**13**	**8**	**10**
25-29	**419**	26	26	36	48	64	63	82	74
30-34	**902**	31	67	84	117	131	181	160	131
35-39	**832**	16	58	78	100	148	154	135	143
40 and over	**310**	14	32	42	41	66	39	39	37

Table 6.1 Maternities with multiple births: occurrence
within/outside marriage and age of mother, 1995-2005
a. numbers

Year	Age of mother at birth							
	All ages	Under 20	20-24	25-29	30-34	35-39	40-44	45 and over
All maternities with multiple births								
1995	**9,038**	291	1,306	2,906	2,993	1,348	175	19
1996	**8,883**	273	1,152	2,643	3,192	1,408	194	21
1997	**9,217**	263	1,124	2,615	3,360	1,599	233	23
1998	**9,080**	303	1,078	2,457	3,332	1,627	252	31
1999	**8,907**	291	958	2,412	3,250	1,728	239	29
2000	**8,792**	297	962	2,163	3,246	1,835	264	25
2001	**8,700**	278	1,003	2,054	3,256	1,785	285	39
2002	**8,861**	301	1,095	1,987	3,165	1,952	314	47
2003	**9,131**	294	1,071	2,006	3,282	2,068	348	62
2004	**9,521**	314	1,138	2,081	3,377	2,157	403	51
2005	**9,543**	303	1,129	2,199	3,267	2,145	441	59
Maternities within marriage with multiple births								
1995	**6,498**	50	623	2,146	2,437	1,100	126	16
1996	**6,323**	28	501	1,914	2,590	1,120	152	18
1997	**6,369**	31	458	1,813	2,653	1,227	174	13
1998	**6,227**	41	427	1,668	2,606	1,269	192	24
1999	**6,011**	41	397	1,593	2,466	1,322	169	23
2000	**5,980**	35	347	1,432	2,539	1,408	195	24
2001	**5,837**	37	375	1,335	2,486	1,376	199	29
2002	**5,929**	26	431	1,327	2,437	1,453	220	35
2003	**5,987**	28	390	1,278	2,498	1,487	261	45
2004	**6,014**	26	354	1,284	2,492	1,533	284	41
2005	**6,000**	25	365	1,328	2,404	1,540	298	40
Maternities outside marriage with multiple births								
1995	**2,540**	241	683	760	556	248	49	3
1996	**2,560**	245	651	729	602	288	42	3
1997	**2,848**	232	666	802	707	372	59	10
1998	**2,853**	262	651	789	726	358	60	7
1999	**2,896**	250	561	819	784	406	70	6
2000	**2,812**	262	615	731	707	427	69	1
2001	**2,863**	241	628	719	770	409	86	10
2002	**2,932**	275	664	660	728	499	94	12
2003	**3,144**	266	681	728	784	581	87	17
2004	**3,507**	288	784	797	885	624	119	10
2005	**3,543**	278	764	871	863	605	143	19

Note: The figures include maternities where live births and/or stillbirths occurred.

Table 6.1   Maternities with multiple births: occurrence within/outside marriage and age of mother, 1995-2005   b. rates[1]

Year	Age of mother at birth							
	All ages	Under 20	20-24	25-29	30-34	35-39	40-44	45 and over
All maternities with multiple births per 1,000 all maternities								
1995	**14.1**	6.9	10.0	13.5	16.7	20.9	16.3	*36.5*
1996	**13.8**	6.1	9.2	12.6	17.4	20.6	17.0	36.9
1997	**14.5**	5.7	9.5	13.0	18.2	21.7	19.1	40.6
1998	**14.4**	6.3	9.5	12.8	17.9	21.0	19.7	56.7
1999	**14.5**	6.0	8.7	13.4	17.8	21.6	17.7	47.4
2000	**14.7**	6.5	9.0	12.8	18.3	22.0	18.5	38.8
2001	**14.8**	6.3	9.3	12.9	18.5	21.0	18.6	53.1
2002	**15.0**	6.9	9.9	13.1	17.8	21.9	19.3	55.3
2003	**14.8**	6.6	9.2	12.9	17.7	21.6	19.3	75.7
2004	**15.0**	7.0	9.4	13.1	18.0	21.4	20.5	58.5
2005	**14.9**	6.8	9.3	13.5	17.6	20.9	21.1	56.5
Maternities within marriage with multiple births per 1,000 maternities within marriage								
1995	**15.3**	8.9	10.3	13.7	17.1	21.9	15.9	*40.9*
1996	**15.3**	5.2	9.2	13.0	18.0	21.4	18.2	*41.3*
1997	**15.9**	5.9	9.4	13.1	18.5	22.0	20.0	*31.0*
1998	**15.9**	7.8	9.4	12.9	18.3	21.8	21.2	58.4
1999	**16.0**	7.7	9.2	13.3	17.8	22.2	18.0	*49.9*
2000	**16.5**	7.4	8.6	12.9	18.9	22.9	19.9	52.2
2001	**16.6**	8.0	9.2	13.1	18.9	22.2	19.0	54.6
2002	**16.9**	5.7	10.6	13.7	18.4	22.6	20.0	57.7
2003	**16.6**	6.4	9.6	13.1	18.3	21.7	21.8	78.5
2004	**16.4**	6.4	8.6	13.1	18.1	21.5	22.0	69.5
2005	**16.4**	6.8	9.2	13.4	17.7	21.2	22.0	56.7
Maternities outside marriage with multiple births per 1,000 maternities outside marriage								
1995	**11.6**	6.6	9.8	12.9	15.2	17.4	17.4	*23.3*
1996	**11.1**	6.2	9.2	11.8	15.0	17.9	13.6	*22.6*
1997	**12.0**	5.6	9.6	12.7	16.9	20.7	16.8	*68.5*
1998	**11.9**	6.1	9.6	12.7	16.7	18.5	16.0	*51.5*
1999	**12.0**	5.8	8.3	13.5	17.6	19.8	17.0	*39.7*
2000	**11.9**	6.4	9.2	12.5	16.3	19.4	15.5	*5.4*
2001	**12.1**	6.1	9.3	12.8	17.2	17.8	17.7	*49.0*
2002	**12.2**	7.1	9.5	11.9	15.8	20.2	17.8	*49.4*
2003	**12.3**	6.7	9.0	12.6	16.1	21.2	14.4	*69.1*
2004	**13.1**	7.0	9.9	13.1	17.7	21.3	17.6	*35.5*
2005	**12.9**	6.8	9.3	13.6	17.2	20.2	19.6	*56.2*

Note: The figures include maternities where live births and/or stillbirths occurred.
1 See section 2.11.

Table 6.2 Maternities with multiple births: whether live births or stillbirths, multiplicity, age of mother and sex, 2005

England and Wales

Age of mother at birth	Maternities	Births					
		Live			Still		
		Total	Male	Female	Total	Male	Female
All maternities with multiple births[1]							
All ages	9,543	18,973	9,602	9,371	261	133	128
Under 20	303	595	303	292	11	7	4
20-24	1,129	2,232	1,115	1,117	37	15	22
25-29	2,199	4,368	2,180	2,188	66	29	37
30-34	3,267	6,508	3,378	3,130	71	42	29
35-39	2,145	4,270	2,129	2,141	62	33	29
40-44	441	881	429	452	13	7	6
45 and over	59	119	68	51	1	-	1
Twins only							
All ages	9,396	18,544	9,394	9,150	248	124	124
Under 20	301	589	303	286	11	7	4
20-24	1,117	2,195	1,102	1,093	37	15	22
25-29	2,164	4,267	2,130	2,137	62	26	36
30-34	3,224	6,383	3,320	3,063	67	40	27
35-39	2,105	4,154	2,065	2,089	58	29	29
40-44	428	843	411	432	12	7	5
45 and over	57	113	63	50	1	-	1
Triplets only							
All ages	146	425	204	221	13	9	4
Under 20	2	6	-	6	-	-	-
20-24	11	33	9	24	-	-	-
25-29	35	101	50	51	4	3	1
30-34	43	125	58	67	4	2	2
35-39	40	116	64	52	4	4	-
40-44	13	38	18	20	1	-	1
45 and over	2	6	5	1	-	-	-

1 Includes quads and above.

Table 6.3 Maternities within marriage with multiple births (numbers and rates[1]): age of mother and number of previous live-born children, 2005

England and Wales

Number of previous live-born children	Age of mother at birth							
	All ages	Under 20	20-24	25-29	30-34	35-39	40-44	45 and over
	Numbers							
Total	6,000	25	365	1,328	2,404	1,540	298	40
0	2,893	18	199	691	1,200	618	139	28
1	1,917	7	118	405	763	531	88	5
2	743	-	40	150	269	245	34	5
3	278	-	6	58	117	78	19	-
4 and over	169	-	2	24	55	68	18	2
	Rates: maternities with multiple births per 1,000 maternities within marriage							
Total	16.4	6.8	9.2	13.4	17.7	21.2	22.0	56.7
0	18.8	6.1	9.0	14.0	21.9	29.6	40.6	133.3
1	14.7	11.2	9.0	12.7	14.6	18.9	19.4	30.5
2	14.4	-	10.7	12.3	14.6	17.1	12.3	39.7
3	15.5	-	7.8	14.5	18.5	14.5	13.6	-
4 and over	15.4	-	13.6	15.6	14.5	17.3	12.7	16.0

Note: The figures include maternities where live births and/or stillbirths occurred.
1 See section 2.11.

Table 6.4   All maternities[1]: age of mother, multiplicity and type of outcome, 2005   **England and Wales**

Outcome	All maternities							
	All ages	Under 20	20-24	25-29	Under 30	30-34	35-39	40 and over
All maternities	**639,627**	**44,813**	**121,644**	**162,986**	**329,443**	**185,749**	**102,536**	**21,899**
Singleton maternities	**630,084**	44,510	120,515	160,787	**325,812**	182,482	100,391	21,399
1 LM	**320,998**	22,534	61,357	81,866	**165,757**	93,096	51,345	10,800
1 LF	**305,864**	21,701	58,556	78,114	**158,371**	88,549	48,498	10,446
1 SM	**1,663**	149	313	424	**886**	406	296	75
1 SF	**1,559**	126	289	383	**798**	431	252	78
All multiple maternities	**9,543**	**303**	**1,129**	**2,199**	**3,631**	**3,267**	**2,145**	**500**
Twins[1]	**9,396**	**301**	**1,117**	**2,164**	**3,582**	**3,224**	**2,105**	**485**
2 LM	**3,093**	122	401	729	**1,252**	1,068	625	148
1 LM and 1 LF	**3,114**	56	289	652	**997**	1,155	788	174
2 LF	**2,979**	114	392	734	**1,240**	947	641	151
1 LM and 1 SM	**72**	5	9	15	**29**	21	20	2
1 LM and 1 SF	**22**	-	2	4	**6**	7	7	2
1 LF and 1 SM	**17**	-	2	1	**3**	4	7	3
1 LF and 1 SF	**61**	2	18	17	**37**	10	11	3
2 SM	**17**	*	*	*	*	*	*	*
1 SM and 1 SF	**1**	*	*	*	*	*	*	*
2 SF	**20**	1	2	7	**10**	4	5	1
Triplets and above [1]	**147**	*	*	*	**49**	**43**	**40**	**15**
3 LM	**26**	*	*	*	7	8	8	3
2 LM and 1 LF	**39**	*	*	*	11	9	14	5
1 LM and 2 LF	**39**	*	*	*	14	12	9	4
3 LF	**31**	*	*	*	13	10	6	2
2 LM and 1 SM	**2**	*	*	*	*	*	*	*
2 LM and 1 SF	-	-	-	-	-	-	-	-
1 LM, 1 LF and 1 SM	**3**	*	*	*	*	*	*	*
1 LM, 1 LF and 1 SF	**1**	*	*	*	*	*	*	*
2 LF and 1 SM	-	-	-	-	-	-	-	-
2 LF and 1 SF	**3**	*	*	*	*	*	*	*
1 LM and 2 SM	**1**	*	*	*	*	*	*	*
1 LM, 1 SM and 1 SF	-	-	-	-	-	-	-	-
1 LM and 2 SF	-	-	-	-	-	-	-	-
1 LF and 2 SM	**1**	*	*	*	*	*	*	*
1 LF, 1 SM and 1 SF	-	-	-	-	-	-	-	-
1 LF and 2 SF	-	-	-	-	-	-	-	-
3 SM	-	-	-	-	-	-	-	-
2 SM and 1 SF	-	-	-	-	-	-	-	-
1 SM and 2 SF	-	-	-	-	-	-	-	-
3 SF	-	-	-	-	-	-	-	-
4 LM	**1**	*	*	*	*	*	*	*

1 Some cell counts have been suppressed to protect the confidentiality of individuals - see section 2.15.
LM - Live-born male   SM - Stillborn male   LF - Live-born female   SF - Stillborn female

Table 7.1 Live births (numbers, rates, general fertility rate, and total fertility rate), and population: occurrence within/outside marriage, and area of usual residence, 2005

England and Wales, Government Office Regions (within England), counties and unitary authorities

Area of usual residence of mother	Estimated number of women aged 15-44 (000s)	Live births			Live births outside marriage per 1,000 live births	General fertility rate (GFR)[1]	Total fertility rate (TFR)[1]
		Total	Within marriage	Outside marriage			
ENGLAND AND WALES	11,057.5	645,835	369,330	276,505	428	58.4	1.80
ENGLAND	10,477.6	613,028	353,667	259,361	423	58.5	1.80
NORTH EAST	522.2	28,249	12,701	15,548	550	54.1	1.72
Darlington UA	19.7	1,218	568	650	534	61.9	2.02
Hartlepool UA	18.2	1,116	403	713	639	61.4	2.03
Middlesbrough UA	29.2	1,916	815	1,101	575	65.7	2.07
Redcar and Cleveland UA	27.0	1,577	598	979	621	58.5	1.95
Stockton-on-Tees UA	38.8	2,241	1,056	1,185	529	57.8	1.87
Durham	100.1	5,189	2,293	2,896	558	51.8	1.68
Northumberland	56.3	3,024	1,542	1,482	490	53.7	1.82
Tyne and Wear (Met County)	232.9	11,968	5,426	6,542	547	51.4	1.59
NORTH WEST	1,410.9	81,722	41,335	40,387	494	57.9	1.82
Blackburn with Darwen UA	29.6	2,284	1,419	865	379	77.2	2.44
Blackpool UA	27.1	1,649	592	1,057	641	60.7	1.98
Halton UA	25.1	1,650	670	980	594	65.8	2.08
Warrington UA	40.2	2,207	1,190	1,017	461	54.9	1.77
Cheshire	131.1	7,317	4,334	2,983	408	55.8	1.81
Cumbria	92.3	4,784	2,535	2,249	470	51.9	1.70
Greater Manchester (Met County)	548.8	33,544	17,210	16,334	487	61.1	1.86
Lancashire	229.6	13,084	7,076	6,008	459	57.0	1.84
Merseyside (Met County)	287.2	15,203	6,309	8,894	585	52.9	1.68
YORKSHIRE AND THE HUMBER	1,041.9	60,665	32,389	28,276	466	58.2	1.83
East Riding of Yorkshire UA	58.0	2,887	1,663	1,224	424	49.7	1.70
Kingston upon Hull, City of UA	54.2	3,203	1,096	2,107	658	59.1	1.77
North East Lincolnshire UA	31.4	1,927	730	1,197	621	61.4	2.05
North Lincolnshire UA	29.6	1,796	873	923	514	60.7	2.05
York UA	41.9	1,929	1,118	811	420	46.1	1.44
North Yorkshire	104.9	5,700	3,526	2,174	381	54.3	1.86
South Yorkshire (Met County)	265.9	15,081	7,278	7,803	517	56.7	1.78
West Yorkshire (Met County)	456.1	28,142	16,105	12,037	428	61.7	1.90
EAST MIDLANDS	868.6	49,080	26,844	22,236	453	56.5	1.80
Derby UA	50.1	2,989	1,618	1,371	459	59.6	1.81
Leicester UA	69.1	4,597	2,734	1,863	405	66.5	1.93
Nottingham UA	69.9	3,753	1,590	2,163	576	53.7	1.60
Rutland UA	6.3	344	239	105	305	54.3	2.33
Derbyshire	142.8	7,902	4,250	3,652	462	55.3	1.83
Leicestershire	123.4	6,657	4,086	2,571	386	53.9	1.75
Lincolnshire	124.2	6,671	3,494	3,177	476	53.7	1.80
Northamptonshire	132.5	8,279	4,622	3,657	442	62.5	2.02
Nottinghamshire	150.1	7,888	4,211	3,677	466	52.5	1.69
WEST MIDLANDS	1,087.6	65,956	37,134	28,822	437	60.6	1.90
Herefordshire, County of UA	31.3	1,654	950	704	426	52.8	1.77
Stoke-on-Trent UA	49.9	3,312	1,473	1,839	555	66.4	2.05
Telford and Wrekin UA	34.1	2,057	967	1,090	530	60.3	1.95
Shropshire	50.8	2,767	1,658	1,109	401	54.5	1.82
Staffordshire	156.7	8,503	4,761	3,742	440	54.2	1.79
Warwickshire	107.2	5,584	3,289	2,295	411	52.1	1.61
West Midlands (Met County)	551.6	36,038	20,529	15,509	430	65.3	1.99
Worcestershire	105.9	6,041	3,507	2,534	419	57.0	1.83

Note: Population estimates may not add exactly due to rounding - see section 2.15.
1 See sections 2.1 and 2.11.

Table 7.1 - *continued*

Area of usual residence of mother	Estimated number of women aged 15-44 (000s)	Live births			Live births outside marriage per 1,000 live births	General fertility rate (GFR)[1]	Total fertility rate (TFR)[1]
		Total	Within marriage	Outside marriage			
EAST	**1,097.9**	**64,687**	**38,996**	**25,691**	**397**	**58.9**	**1.84**
Luton UA	40.6	**3,196**	2,180	1,016	318	78.7	2.34
Peterborough UA	33.7	**2,446**	1,394	1,052	430	72.5	2.25
Southend-on-Sea UA	30.9	**1,948**	997	951	488	63.0	1.98
Thurrock UA	32.4	**2,211**	1,158	1,053	476	68.3	2.08
Bedfordshire	82.0	**4,881**	3,027	1,854	380	59.6	1.86
Cambridgeshire	124.2	**6,505**	4,209	2,296	353	52.4	1.58
Essex	259.1	**14,905**	8,921	5,984	401	57.5	1.83
Hertfordshire	217.1	**12,975**	8,424	4,551	351	59.8	1.83
Norfolk	150.3	**8,127**	4,300	3,827	471	54.1	1.72
Suffolk	127.6	**7,493**	4,386	3,107	415	58.7	1.90
LONDON	**1,849.3**	**116,019**	**75,779**	**40,240**	**347**	**62.7**	**1.77**
Inner London	817.7	**49,548**	31,681	17,867	361	60.6	1.67
Outer London	1,031.7	**66,471**	44,098	22,373	337	64.4	1.88
SOUTH EAST	**1,640.3**	**93,921**	**58,319**	**35,602**	**379**	**57.3**	**1.78**
Bracknell Forest UA	25.1	**1,431**	935	496	347	56.9	1.73
Brighton and Hove UA	62.2	**3,045**	1,608	1,437	472	49.0	1.40
Isle of Wight UA	23.7	**1,188**	582	606	510	50.1	1.69
Medway UA	53.6	**3,134**	1,630	1,504	480	58.5	1.85
Milton Keynes UA	47.5	**3,200**	1,818	1,382	432	67.4	2.07
Portsmouth UA	44.0	**2,333**	1,172	1,161	498	53.0	1.58
Reading UA	35.0	**2,146**	1,265	881	411	61.4	1.74
Slough UA	27.3	**2,108**	1,452	656	311	77.3	2.25
Southampton UA	52.7	**2,775**	1,415	1,360	490	52.6	1.52
West Berkshire UA	29.3	**1,796**	1,207	589	328	61.2	1.98
Windsor and Maidenhead UA	28.4	**1,699**	1,238	461	271	59.7	1.77
Wokingham UA	32.2	**1,694**	1,284	410	242	52.7	1.61
Buckinghamshire	94.8	**5,511**	3,922	1,589	288	58.1	1.82
East Sussex	83.8	**4,783**	2,633	2,150	450	57.1	1.95
Hampshire	243.5	**13,537**	8,699	4,838	357	55.6	1.79
Kent	267.1	**15,613**	8,655	6,958	446	58.5	1.88
Oxfordshire	134.6	**7,575**	5,102	2,473	326	56.3	1.70
Surrey	216.1	**12,303**	8,702	3,601	293	56.9	1.73
West Sussex	139.4	**8,050**	5,000	3,050	379	57.7	1.85
SOUTH WEST	**958.9**	**52,729**	**30,170**	**22,559**	**428**	**55.0**	**1.76**
Bath and North East Somerset UA	35.8	**1,702**	1,031	671	394	47.5	1.53
Bournemouth UA	34.2	**1,619**	898	721	445	47.4	1.40
Bristol, City of UA	95.9	**5,440**	2,859	2,581	474	56.7	1.65
North Somerset UA	34.5	**2,032**	1,196	836	411	59.0	1.96
Plymouth UA	51.9	**2,813**	1,376	1,437	511	54.2	1.70
Poole UA	25.6	**1,461**	867	594	407	57.0	1.85
South Gloucestershire UA	49.9	**2,955**	1,883	1,072	363	59.2	1.88
Swindon UA	39.3	**2,480**	1,382	1,098	443	63.2	1.96
Torbay UA	22.7	**1,328**	647	681	513	58.4	1.97
Cornwall and Isles of Scilly	90.8	**4,789**	2,520	2,269	474	52.7	1.75
Devon	128.1	**6,693**	3,898	2,795	418	52.2	1.74
Dorset	63.8	**3,487**	2,035	1,452	416	54.7	1.94
Gloucestershire	110.4	**5,946**	3,524	2,422	407	53.9	1.75
Somerset	92.0	**5,149**	2,877	2,272	441	56.0	1.90
Wiltshire	84.0	**4,835**	3,177	1,658	343	57.5	1.91

1 See sections 2.1 and 2.11.

Table 7.1 - *continued*

Area of usual residence of mother	Estimated number of women aged 15-44 (000s)	Live births			Live births outside marriage per 1,000 live births	General fertility rate (GFR)[1]	Total fertility rate (TFR)[1]
		Total	Within marriage	Outside marriage			
WALES	**579.9**	**32,593**	**15,518**	**17,075**	**524**	**56.2**	**1.79**
Isle of Anglesey	12.0	**685**	326	359	524	56.9	1.88
Gwynedd	22.0	**1,264**	512	752	595	57.5	1.81
Conwy	18.7	**1,044**	491	553	530	55.9	1.87
Denbighshire	17.1	**984**	462	522	530	57.4	1.93
Flintshire	29.5	**1,642**	863	779	474	55.6	1.80
Wrexham	25.9	**1,582**	743	839	530	61.1	1.92
Powys	21.9	**1,248**	672	576	462	57.0	1.94
Ceredigion	15.5	**600**	321	279	465	38.8	1.40
Pembrokeshire	20.5	**1,201**	570	631	525	58.5	1.97
Carmarthenshire	32.3	**1,744**	889	855	490	54.0	1.78
Swansea	45.0	**2,449**	1,161	1,288	526	54.4	1.72
Neath Port Talbot	25.8	**1,486**	670	816	549	57.7	1.92
Bridgend	25.6	**1,523**	739	784	515	59.5	1.97
The Vale of Glamorgan	23.9	**1,279**	691	588	460	53.4	1.76
Cardiff	77.1	**3,955**	2,081	1,874	474	51.3	1.54
Rhondda, Cynon, Taff	47.2	**2,864**	1,184	1,680	587	60.7	1.93
Merthyr Tydfil	11.0	**643**	223	420	653	58.6	1.96
Caerphilly	34.2	**2,055**	897	1,158	564	60.1	1.93
Blaenau Gwent	13.6	**734**	252	482	657	54.1	1.84
Torfaen	17.5	**1,094**	439	655	599	62.4	2.07
Monmouthshire	15.3	**819**	503	316	386	53.4	1.87
Newport	28.3	**1,698**	829	869	512	60.0	1.97
Normal residence outside England and Wales	:	**214**	145	69	322	:	:

Note: Population estimates may not add exactly due to rounding - see section 2.15.
1 See sections 2.1 and 2.11.

Table 7.2   Live births (numbers, rates, general fertility rate, and total fertility rate), and population: occurrence within/outside marriage, and area of usual residence, 2005

Area of usual residence of mother	Estimated number of women aged 15-44 (000s)	Live births			Live births outside marriage per 1,000 live births	General fertility rate (GFR)[1]	Total fertility rate (TFR)[1]
		Total	Within marriage	Outside marriage			
ENGLAND AND WALES	**11,057.5**	**645,835**	**369,330**	**276,505**	**428**	**58.4**	**1.80**
ENGLAND	**10,477.6**	**613,028**	**353,667**	**259,361**	**423**	**58.5**	**1.80**
NORTH EAST	**522.2**	**28,249**	**12,701**	**15,548**	**550**	**54.1**	**1.72**
County Durham and Tees Valley	232.9	**13,257**	5,733	7,524	568	56.9	1.85
Northumberland, Tyne & Wear	289.2	**14,992**	6,968	8,024	535	51.8	1.63
NORTH WEST	**1,410.9**	**81,722**	**41,335**	**40,387**	**494**	**57.9**	**1.82**
Cheshire & Merseyside	483.5	**26,377**	12,503	13,874	526	54.6	1.75
Cumbria and Lancashire	378.6	**21,801**	11,622	10,179	467	57.6	1.87
Greater Manchester	548.8	**33,544**	17,210	16,334	487	61.1	1.86
YORKSHIRE AND THE HUMBER	**1,041.9**	**60,665**	**32,389**	**28,276**	**466**	**58.2**	**1.83**
North and East Yorkshire and Northern Lincolnshire	319.9	**17,442**	9,006	8,436	484	54.5	1.78
South Yorkshire	265.9	**15,081**	7,278	7,803	517	56.7	1.78
West Yorkshire	456.1	**28,142**	16,105	12,037	428	61.7	1.90
EAST MIDLANDS	**868.6**	**49,080**	**26,844**	**22,236**	**453**	**56.5**	**1.80**
Leicestershire, Northamptonshire and Rutland	331.4	**19,877**	11,681	8,196	412	60.0	1.90
Trent	537.2	**29,203**	15,163	14,040	481	54.4	1.73
WEST MIDLANDS	**1,087.6**	**65,956**	**37,134**	**28,822**	**437**	**60.6**	**1.90**
Birmingham and the Black Country	486.0	**32,167**	18,475	13,692	426	66.2	2.02
Shropshire and Staffordshire	291.5	**16,639**	8,859	7,780	468	57.1	1.87
West Midlands South	310.0	**17,150**	9,800	7,350	429	55.3	1.74
EAST	**1,097.9**	**64,687**	**38,996**	**25,691**	**397**	**58.9**	**1.84**
Bedfordshire and Hertfordshire	339.7	**21,052**	13,631	7,421	353	62.0	1.91
Essex	322.4	**19,064**	11,076	7,988	419	59.1	1.87
Norfolk, Suffolk and Cambridgeshire	435.9	**24,571**	14,289	10,282	418	56.4	1.77
LONDON	**1,849.3**	**116,019**	**75,779**	**40,240**	**347**	**62.7**	**1.77**
North Central London	313.5	**18,935**	12,545	6,390	337	60.4	1.70
North East London	370.7	**26,785**	17,714	9,071	339	72.3	2.06
North West London	468.6	**27,162**	19,629	7,533	277	58.0	1.61
South East London	367.0	**24,049**	13,027	11,022	458	65.5	1.89
South West London	329.6	**19,088**	12,864	6,224	326	57.9	1.60
SOUTH EAST	**1,640.3**	**93,921**	**58,319**	**35,602**	**379**	**57.3**	**1.78**
Hampshire and Isle of Wight	364.0	**19,833**	11,868	7,965	402	54.5	1.71
Kent and Medway	320.6	**18,747**	10,285	8,462	451	58.5	1.88
Surrey and Sussex	501.5	**28,181**	17,943	10,238	363	56.2	1.74
Thames Valley	454.2	**27,160**	18,223	8,937	329	59.8	1.81
SOUTH WEST	**958.9**	**52,729**	**30,170**	**22,559**	**428**	**55.0**	**1.76**
Avon, Gloucestershire and Wiltshire	449.8	**25,390**	15,052	10,338	407	56.4	1.76
Dorset and Somerset	215.6	**11,716**	6,677	5,039	430	54.4	1.79
South West Peninsula	293.5	**15,623**	8,441	7,182	460	53.2	1.75

Note: Population estimates may not add exactly due to rounding - see section 2.15.
1 See sections 2.1 and 2.11.

Table 7.2 - *continued*

Area of usual residence of mother	Estimated number of women aged 15-44 (000s)	Live births			Live births outside marriage per 1,000 live births	General fertility rate (GFR)[1]	Total fertility rate (TFR)[1]
		Total	Within marriage	Outside marriage			
WALES	**579.9**	**32,593**	**15,518**	**17,075**	**524**	**56.2**	**1.79**
Anglesey	12.0	685	326	359	524	56.9	1.88
Gwynedd	22.0	1,264	512	752	595	57.5	1.81
Conwy	18.7	1,044	491	553	530	55.9	1.87
Denbighshire	17.1	984	462	522	530	57.4	1.93
Flintshire	29.5	1,642	863	779	474	55.6	1.80
Wrexham	25.9	1,582	743	839	530	61.1	1.92
Powys	21.9	1,248	672	576	462	57.0	1.94
Ceredigion	15.5	600	321	279	465	38.8	1.40
Pembrokeshire	20.5	1,201	570	631	525	58.5	1.97
Carmarthenshire	32.3	1,744	889	855	490	54.0	1.78
Swansea	45.0	2,449	1,161	1,288	526	54.4	1.72
Neath Port Talbot	25.8	1,486	670	816	549	57.7	1.92
Bridgend	25.6	1,523	739	784	515	59.5	1.97
Vale of Glamorgan	23.9	1,279	691	588	460	53.4	1.76
Cardiff	77.1	3,955	2,081	1,874	474	51.3	1.54
Rhondda Cynon Taff	47.2	2,864	1,184	1,680	587	60.7	1.93
Merthyr Tydfil	11.0	643	223	420	653	58.6	1.96
Caerphilly	34.2	2,055	897	1,158	564	60.1	1.93
Blaenau Gwent	13.6	734	252	482	657	54.1	1.84
Torfaen	17.5	1,094	439	655	599	62.4	2.07
Monmouthshire	15.3	819	503	316	386	53.4	1.87
Newport	28.3	1,698	829	869	512	60.0	1.97
Normal residence outside England and Wales	:	214	145	69	322	:	:

Note: Population estimates may not add due to rounding - see section 2.15.
1 See sections 2.1 and 2.11.

Table 7.3   Live births (numbers and rates): age of
mother and area of usual residence, 2005

**England and Wales, Government
Office Regions (within England),
counties and unitary authorities**

Area of usual residence of mother	Age of mother at birth							Age of mother at birth						
	All ages	Under 20	20-24	25-29	30-34	35-39	40 and over	All ages[1]	Under 20	20-24	25-29	30-34	35-39	40 and over
	Numbers							**Rates per 1,000 women in age group[2]**						
ENGLAND AND WALES	645,835	44,830	122,145	164,348	188,153	104,113	22,246	58.4	26.3	71.7	98.8	100.9	50.3	10.8
ENGLAND	613,028	41,723	114,882	155,897	179,541	99,663	21,322	58.5	26.0	71.4	98.5	101.3	50.8	11.0
NORTH EAST	28,249	2,849	6,646	7,300	7,245	3,527	682	54.1	33.4	75.8	100.4	89.0	37.1	6.8
Darlington UA	1,218	103	268	352	341	127	27	61.9	33.1	104.9	120.5	100.4	33.8	6.8
Hartlepool UA	1,116	148	329	273	230	122	14	61.4	46.5	121.9	109.2	84.6	34.8	3.9
Middlesbrough UA	1,916	206	545	532	389	208	36	65.7	38.7	101.3	131.1	92.8	42.4	6.8
Redcar and Cleveland UA	1,577	186	392	407	381	180	31	58.5	38.9	98.3	117.0	92.8	34.5	5.7
Stockton-on-Tees UA	2,241	211	526	586	583	285	50	57.8	33.4	90.4	108.2	93.6	39.2	6.5
Durham	5,189	525	1,208	1,310	1,356	667	123	51.8	31.9	72.0	101.7	86.5	35.2	6.4
Northumberland	3,024	263	632	741	844	431	113	53.7	28.7	82.8	105.2	94.4	38.3	9.2
Tyne and Wear (Met County)	11,968	1,207	2,746	3,099	3,121	1,507	288	51.4	32.6	64.1	90.0	86.4	37.4	6.8
NORTH WEST	81,722	6,875	17,480	20,915	22,103	11,977	2,372	57.9	29.8	77.3	104.6	96.6	45.9	9.0
Blackburn with Darwen UA	2,284	194	637	648	513	244	48	77.2	35.9	145.2	144.7	104.1	46.2	9.4
Blackpool UA	1,649	224	417	414	356	192	46	60.7	50.2	102.7	117.5	80.5	35.6	8.7
Halton UA	1,650	151	389	432	424	224	30	65.8	35.8	99.7	120.6	101.9	50.3	6.3
Warrington UA	2,207	154	369	554	667	397	66	54.9	25.0	67.8	104.3	99.3	48.3	7.9
Cheshire	7,317	447	1,107	1,661	2,335	1,473	294	55.8	21.4	62.8	102.0	106.9	54.8	10.6
Cumbria	4,784	327	979	1,152	1,392	767	167	51.9	21.8	77.7	96.8	91.3	41.6	8.8
Greater Manchester (Met County)	33,544	2,999	7,601	8,757	8,803	4,532	852	61.1	34.5	83.0	102.9	97.1	46.0	8.9
Lancashire	13,084	1,097	2,725	3,449	3,553	1,900	360	57.0	27.9	74.8	114.2	96.7	44.4	8.1
Merseyside (Met County)	15,203	1,282	3,256	3,848	4,060	2,248	509	52.9	26.3	65.1	97.5	92.1	44.2	9.4
YORKSHIRE AND THE HUMBER	60,665	5,482	13,540	15,941	16,070	8,039	1,593	58.2	32.1	78.2	109.3	95.4	42.0	8.3
East Riding of Yorkshire UA	2,887	187	462	681	976	495	86	49.7	18.6	64.5	103.3	102.6	41.1	6.8
Kingston upon Hull, City of UA	3,203	456	944	839	640	275	49	59.1	50.6	91.1	101.1	75.2	30.9	5.4
North East Lincolnshire UA	1,927	260	555	494	381	207	30	61.4	45.2	121.5	122.2	78.5	35.0	4.8
North Lincolnshire UA	1,796	174	387	501	440	241	53	60.7	35.7	97.5	132.6	92.0	39.7	8.7
York UA	1,929	125	301	467	593	365	78	46.1	19.1	33.3	75.9	93.6	52.5	11.4
North Yorkshire	5,700	343	921	1,348	1,784	1,041	263	54.3	19.2	74.3	108.5	104.5	48.3	11.2
South Yorkshire (Met County)	15,081	1,522	3,428	3,930	3,846	1,962	393	56.7	35.5	75.2	106.1	90.1	39.8	8.1
West Yorkshire (Met County)	28,142	2,415	6,542	7,681	7,410	3,453	641	61.7	32.8	81.7	113.8	99.2	42.7	8.1
EAST MIDLANDS	49,080	3,835	9,711	12,604	14,103	7,346	1,481	56.5	27.5	71.9	106.0	98.8	44.1	8.9
Derby UA	2,989	269	660	819	771	392	78	59.6	34.9	73.8	106.8	93.4	43.3	9.2
Leicester UA	4,597	379	1,122	1,362	1,141	482	111	66.5	34.7	72.4	118.0	105.0	45.7	11.4
Nottingham UA	3,753	469	1,001	920	838	433	92	53.7	40.7	53.0	84.7	85.1	43.9	10.4
Rutland UA	344	15	60	75	123	53	18	54.3	10.2	133.6	105.2	116.9	40.7	13.3
Derbyshire	7,902	586	1,480	1,912	2,350	1,309	265	55.3	26.3	84.7	104.1	94.5	43.8	8.8
Leicestershire	6,657	350	1,029	1,714	2,192	1,163	209	53.9	17.5	57.9	100.9	106.8	47.0	8.4
Lincolnshire	6,671	573	1,374	1,760	1,803	973	188	53.7	27.1	77.7	113.0	91.1	39.8	7.4
Northamptonshire	8,279	582	1,583	2,148	2,484	1,231	251	62.5	27.7	87.8	120.4	109.9	46.0	9.6
Nottinghamshire	7,888	612	1,402	1,894	2,401	1,310	269	52.5	26.0	69.0	91.6	96.1	43.7	8.8
WEST MIDLANDS	65,956	5,404	14,268	17,406	17,702	9,247	1,929	60.6	30.3	83.6	110.6	99.6	45.5	9.6
Herefordshire, County of UA	1,654	119	273	379	491	320	72	52.8	22.1	72.2	100.1	96.4	49.8	10.6
Stoke-on-Trent UA	3,312	402	940	911	699	309	51	66.4	49.0	104.8	129.0	86.0	35.2	5.8
Telford and Wrekin UA	2,057	236	488	499	532	256	46	60.3	41.8	102.7	107.5	89.8	37.9	7.2
Shropshire	2,767	154	464	653	866	527	103	54.5	17.9	73.5	105.4	104.0	50.2	9.5
Staffordshire	8,503	588	1,554	2,171	2,560	1,366	264	54.2	22.5	72.1	110.7	98.3	43.5	8.2
Warwickshire	5,584	362	871	1,398	1,782	954	217	52.1	22.4	54.8	86.6	99.8	46.6	10.5
West Midlands (Met County)	36,038	3,113	8,734	9,896	8,879	4,462	954	65.3	34.2	91.2	116.2	100.4	45.9	10.2
Worcestershire	6,041	430	944	1,499	1,893	1,053	222	57.0	25.5	70.0	101.6	105.8	48.5	10.5

1   The rates for women of all ages, under 20 and 40 and over are based on women aged 15-44, 15-19 and 40-44 respectively.
2   See sections 2.1 and 2.11.

Table 7.3 - *continued*

Area of usual residence of mother	Age of mother at birth							Age of mother at birth						
	All ages	Under 20	20-24	25-29	30-34	35-39	40 and over	All ages[1]	Under 20	20-24	25-29	30-34	35-39	40 and over
	Numbers							**Rates per 1,000 women in age group**[2]						
EAST	**64,687**	**3,586**	**10,995**	**16,826**	**20,224**	**10,866**	**2,190**	**58.9**	**21.1**	**71.2**	**106.1**	**107.6**	**50.9**	**10.3**
Luton UA	**3,196**	198	779	981	797	349	92	**78.7**	30.2	109.2	140.9	121.1	51.4	14.0
Peterborough UA	**2,446**	223	599	712	570	280	62	**72.5**	42.3	119.5	134.3	97.4	44.5	10.3
Southend-on-Sea UA	**1,948**	135	390	492	563	311	57	**63.0**	29.0	93.2	106.2	104.7	52.2	9.4
Thurrock UA	**2,211**	139	455	650	610	309	48	**68.3**	29.1	105.6	120.5	101.6	49.6	8.5
Bedfordshire	**4,881**	253	763	1,319	1,514	854	178	**59.6**	20.7	69.3	112.1	104.4	52.7	11.0
Cambridgeshire	**6,505**	306	971	1,604	2,209	1,166	249	**52.4**	16.4	48.2	85.5	103.9	50.7	11.1
Essex	**14,905**	753	2,266	3,844	4,972	2,618	452	**57.5**	18.4	64.8	115.1	115.1	50.5	8.7
Hertfordshire	**12,975**	538	1,667	3,025	4,513	2,689	543	**59.8**	16.7	60.3	96.2	116.6	61.7	12.4
Norfolk	**8,127**	580	1,647	2,184	2,299	1,154	263	**54.1**	24.2	72.0	104.9	92.2	40.7	9.0
Suffolk	**7,493**	461	1,458	2,015	2,177	1,136	246	**58.7**	22.4	84.7	114.6	100.8	45.1	9.7
LONDON	**116,019**	**5,254**	**18,380**	**29,279**	**35,753**	**21,937**	**5,416**	**62.7**	**23.7**	**64.5**	**80.4**	**101.2**	**66.6**	**18.4**
Inner London	**49,548**	2,304	8,137	11,946	14,986	9,686	2,489	**60.6**	27.5	60.5	64.2	90.6	71.3	22.2
Outer London	**66,471**	2,950	10,243	17,333	20,767	12,251	2,927	**64.4**	21.4	68.0	97.3	110.5	63.3	16.0
SOUTH EAST	**93,921**	**5,132**	**14,578**	**22,709**	**30,118**	**17,654**	**3,730**	**57.3**	**20.3**	**61.6**	**96.1**	**109.4**	**55.6**	**11.6**
Bracknell Forest UA	**1,431**	73	202	344	469	280	63	**56.9**	20.2	64.4	87.4	100.6	55.8	13.2
Brighton and Hove UA	**3,045**	146	378	590	1,056	716	159	**49.0**	18.7	32.2	56.1	93.2	65.3	16.2
Isle of Wight UA	**1,188**	105	260	311	309	155	48	**50.1**	25.7	79.5	102.4	84.9	32.1	9.8
Medway UA	**3,134**	239	704	926	813	382	70	**58.5**	27.9	87.9	115.5	91.4	38.0	6.9
Milton Keynes UA	**3,200**	215	595	914	913	464	99	**67.4**	30.5	94.4	119.8	105.6	49.9	11.6
Portsmouth UA	**2,333**	213	520	664	592	271	73	**53.0**	31.3	56.2	93.0	86.6	38.3	10.6
Reading UA	**2,146**	154	407	562	627	328	68	**61.4**	33.9	56.7	83.2	100.9	59.5	14.3
Slough UA	**2,108**	98	461	696	561	243	49	**77.3**	26.0	121.6	132.3	108.5	50.7	10.9
Southampton UA	**2,775**	247	646	785	726	315	56	**52.6**	31.8	48.6	82.6	88.7	44.3	8.1
West Berkshire UA	**1,796**	82	220	398	676	361	59	**61.2**	16.5	74.2	102.4	127.8	59.3	9.6
Windsor and Maidenhead UA	**1,699**	59	197	324	609	402	108	**59.7**	15.0	56.5	70.7	119.2	71.6	18.9
Wokingham UA	**1,694**	41	150	338	672	414	79	**52.7**	8.1	35.6	75.8	124.7	65.6	11.8
Buckinghamshire	**5,511**	155	694	1,277	1,907	1,209	269	**58.1**	10.4	59.5	99.3	117.8	62.3	13.6
East Sussex	**4,783**	329	816	1,146	1,407	878	207	**57.1**	22.1	78.3	113.5	108.9	52.0	11.1
Hampshire	**13,537**	761	1,987	3,313	4,444	2,515	517	**55.6**	19.7	62.2	103.0	109.7	51.5	10.0
Kent	**15,613**	1,086	2,895	3,946	4,584	2,579	523	**58.5**	24.8	77.8	107.8	104.0	49.2	9.9
Oxfordshire	**7,575**	359	1,062	1,662	2,535	1,631	326	**56.3**	17.9	45.7	83.7	112.8	65.7	13.5
Surrey	**12,303**	361	1,218	2,558	4,591	2,942	633	**56.9**	11.5	43.0	82.6	125.8	67.0	14.1
West Sussex	**8,050**	409	1,166	1,955	2,627	1,569	324	**57.7**	18.7	67.2	103.0	112.7	55.2	11.0
SOUTH WEST	**52,729**	**3,306**	**9,284**	**12,917**	**16,223**	**9,070**	**1,929**	**55.0**	**21.1**	**66.3**	**99.6**	**103.8**	**48.8**	**10.1**
Bath and North East Somerset UA	**1,702**	84	207	384	596	347	84	**47.5**	13.8	29.9	79.4	110.3	55.6	13.3
Bournemouth UA	**1,619**	73	284	426	499	276	61	**47.4**	15.1	42.3	75.1	86.9	48.9	10.9
Bristol, City of UA	**5,440**	396	1,009	1,266	1,666	931	172	**56.7**	30.0	48.6	72.5	104.5	63.2	12.4
North Somerset UA	**2,032**	100	339	460	681	372	80	**59.0**	17.9	80.4	109.3	119.1	51.0	10.7
Plymouth UA	**2,813**	260	642	776	714	360	61	**54.2**	30.6	64.0	108.9	90.2	39.1	6.7
Poole UA	**1,461**	77	234	368	474	255	53	**57.0**	17.6	70.1	107.9	110.7	50.5	10.2
South Gloucestershire UA	**2,955**	160	431	735	1,011	526	92	**59.2**	21.6	68.1	108.6	117.1	50.5	8.9
Swindon UA	**2,480**	186	463	641	770	358	62	**63.2**	33.2	89.9	104.3	107.3	46.9	8.2
Torbay UA	**1,328**	118	279	342	348	203	38	**58.4**	30.4	87.4	121.3	96.8	46.3	7.8
Cornwall and Isles of Scilly	**4,789**	336	925	1,224	1,329	798	177	**52.7**	21.9	76.2	106.2	89.1	43.9	9.5
Devon	**6,693**	406	1,118	1,595	2,074	1,233	267	**52.2**	18.3	59.1	103.3	105.2	48.9	10.0
Dorset	**3,487**	217	617	818	1,085	612	138	**54.7**	18.9	84.9	116.0	107.5	46.3	9.4
Gloucestershire	**5,946**	353	1,015	1,422	1,847	1,064	245	**53.9**	19.2	69.0	96.6	103.7	48.2	10.8
Somerset	**5,149**	317	962	1,295	1,558	828	189	**56.0**	19.8	85.3	110.9	104.4	44.2	9.8
Wiltshire	**4,835**	223	759	1,165	1,571	907	210	**57.5**	16.4	83.3	107.5	109.1	50.3	11.6

1 The rates for women of all ages, under 20 and 40 and over are based on women aged 15-44, 15-19 and 40-44 respectively.
2 See sections 2.1 and 2.11.

Table 7.3 - *continued*

Area of usual residence of mother	Age of mother at birth							Age of mother at birth						
	All ages	Under 20	20-24	25-29	30-34	35-39	40 and over	All ages[1]	Under 20	20-24	25-29	30-34	35-39	40 and over
	Numbers							Rates per 1,000 women in age group[2]						
WALES	**32,593**	**3,099**	**7,228**	**8,393**	**8,553**	**4,416**	**904**	**56.2**	**32.0**	**77.1**	**105.0**	**92.8**	**41.3**	**8.2**
Isle of Anglesey	**685**	63	151	197	166	86	22	**56.9**	30.5	92.8	114.6	84.7	37.7	9.2
Gwynedd	**1,264**	116	259	317	327	197	48	**57.5**	30.5	64.4	107.5	94.5	51.2	12.3
Conwy	**1,044**	95	198	257	289	174	31	**55.9**	30.4	82.7	106.0	97.8	47.0	7.6
Denbighshire	**984**	92	212	256	259	136	29	**57.4**	31.3	100.0	107.2	95.7	40.6	8.0
Flintshire	**1,642**	136	323	408	454	270	51	**55.6**	30.2	82.4	100.0	89.6	45.4	8.5
Wrexham	**1,582**	153	336	400	432	231	30	**61.1**	39.3	83.9	106.1	98.3	46.3	6.2
Powys	**1,248**	77	188	324	367	237	55	**57.0**	20.4	72.3	121.2	103.5	52.9	11.4
Ceredigion	**600**	46	109	146	174	94	31	**38.8**	16.1	26.3	81.8	90.9	42.5	12.1
Pembrokeshire	**1,201**	110	281	313	295	174	28	**58.5**	30.5	103.8	112.8	91.3	45.0	6.4
Carmarthenshire	**1,744**	139	352	481	468	248	56	**54.0**	25.1	77.0	109.7	92.5	40.3	8.5
Swansea	**2,449**	253	569	605	638	315	69	**54.4**	32.8	65.4	100.7	92.1	40.7	8.7
Neath Port Talbot	**1,486**	154	356	411	389	145	31	**57.7**	35.1	94.5	121.3	95.3	29.9	5.8
Bridgend	**1,523**	145	336	404	416	184	38	**59.5**	34.4	98.3	120.4	93.9	36.9	7.3
The Vale of Glamorgan	**1,279**	84	244	324	356	222	49	**53.4**	20.3	76.1	101.2	94.4	47.6	9.9
Cardiff	**3,955**	339	850	917	1,113	607	129	**51.3**	28.9	46.8	73.6	97.3	51.7	11.2
Rhondda, Cynon, Taff	**2,864**	327	723	809	671	289	45	**60.7**	42.0	96.5	119.4	85.5	33.3	5.2
Merthyr Tydfil	**643**	95	217	146	127	46	12	**58.6**	49.5	127.1	109.9	73.7	21.7	5.5
Caerphilly	**2,055**	237	500	559	516	207	36	**60.1**	41.9	102.8	113.2	87.5	31.8	5.7
Blaenau Gwent	**734**	91	234	198	128	75	8	**54.1**	36.4	121.6	112.9	60.5	28.5	3.0
Torfaen	**1,094**	141	280	295	241	116	21	**62.4**	46.0	114.2	123.4	86.6	34.9	6.0
Monmouthshire	**819**	43	114	184	277	158	43	**53.4**	16.2	68.2	107.0	116.5	47.4	12.0
Newport	**1,698**	163	396	442	450	205	42	**60.0**	33.6	93.7	119.6	99.3	36.5	7.8
Normal residence outside England and Wales	**214**	8	35	58	59	34	20	:	:	:	:	:	:	:

1 The rates for women of all ages, under 20 and 40 and over are based on women aged 15-44, 15-19 and 40-44 respectively.
2 See sections 2.1 and 2.11

Table 7.4   Live births (numbers and rates): age of
mother and area of usual residence, 2005

England and Wales, Government Office
Regions (within England), and
health authorities/boards

Area of usual residence of mother	Age of mother at birth							Age of mother at birth						
	All ages	Under 20	20-24	25-29	30-34	35-39	40 and over	All ages[1]	Under 20	20-24	25-29	30-34	35-39	40 and over
	Numbers							Rates per 1,000 women in age group[2]						
ENGLAND AND WALES	645,835	44,830	122,145	164,348	188,153	104,113	22,246	58.4	26.3	71.7	98.8	100.9	50.3	10.8
ENGLAND	613,028	41,723	114,882	155,897	179,541	99,663	21,322	58.5	26.0	71.4	98.5	101.3	50.8	11.0
NORTH EAST	28,249	2,849	6,646	7,300	7,245	3,527	682	54.1	33.4	75.8	100.4	89.0	37.1	6.8
County Durham and Tees Valley	13,257	1,379	3,268	3,460	3,280	1,589	281	56.9	35.2	87.8	110.7	90.3	36.4	6.2
Northumberland, Tyne & Wear	14,992	1,470	3,378	3,840	3,965	1,938	401	51.8	31.8	66.9	92.6	88.0	37.6	7.4
NORTH WEST	81,722	6,875	17,480	20,915	22,103	11,977	2,372	57.9	29.8	77.3	104.6	96.6	45.9	9.0
Cheshire & Merseyside	26,377	2,034	5,121	6,495	7,486	4,342	899	54.6	25.4	66.5	100.5	97.5	48.0	9.5
Cumbria and Lancashire	21,801	1,842	4,758	5,663	5,814	3,103	621	57.6	28.7	82.8	113.1	94.8	43.2	8.4
Greater Manchester	33,544	2,999	7,601	8,757	8,803	4,532	852	61.1	34.5	83.0	102.9	97.1	46.0	8.9
YORKSHIRE AND THE HUMBER	60,665	5,482	13,540	15,941	16,070	8,039	1,593	58.2	32.1	78.2	109.3	95.4	42.0	8.3
North and East Yorkshire and Northern Lincolnshire	17,442	1,545	3,570	4,330	4,814	2,624	559	54.5	28.6	75.2	104.9	94.3	42.7	8.7
South Yorkshire	15,081	1,522	3,428	3,930	3,846	1,962	393	56.7	35.5	75.2	106.1	90.1	39.8	8.1
West Yorkshire	28,142	2,415	6,542	7,681	7,410	3,453	641	61.7	32.8	81.7	113.8	99.2	42.7	8.1
EAST MIDLANDS	49,080	3,835	9,711	12,604	14,103	7,346	1,481	56.5	27.5	71.9	106.0	98.8	44.1	8.9
Leicestershire, Northamptonshire and Rutland	19,877	1,326	3,794	5,299	5,940	2,929	589	60.0	24.8	73.3	115.9	107.9	46.2	9.5
Trent	29,203	2,509	5,917	7,305	8,163	4,417	892	54.4	29.1	71.0	99.9	93.0	42.8	8.6
WEST MIDLANDS	65,956	5,404	14,268	17,406	17,702	9,247	1,929	60.6	30.3	83.6	110.6	99.6	45.5	9.6
Birmingham and the Black Country	32,167	2,773	7,821	8,811	7,890	4,019	853	66.2	34.7	93.9	117.9	100.8	46.6	10.2
Shropshire and Staffordshire	16,639	1,380	3,446	4,234	4,657	2,458	464	57.1	28.4	82.9	112.9	96.2	42.8	8.0
West Midlands South	17,150	1,251	3,001	4,361	5,155	2,770	612	55.3	25.2	65.7	96.7	101.1	46.5	10.4
EAST	64,687	3,586	10,995	16,826	20,224	10,866	2,190	58.9	21.1	71.2	106.1	107.6	50.9	10.3
Bedfordshire and Hertfordshire	21,052	989	3,209	5,325	6,824	3,892	813	62.0	19.4	70.1	106.1	114.1	58.5	12.2
Essex	19,064	1,027	3,111	4,986	6,145	3,238	557	59.1	20.4	71.6	108.4	112.6	50.6	8.7
Norfolk, Suffolk and Cambridgeshire	24,571	1,570	4,675	6,515	7,255	3,736	820	56.4	22.9	71.7	104.3	98.5	45.1	9.9
LONDON	116,019	5,254	18,380	29,279	35,753	21,937	5,416	62.7	23.7	64.5	80.4	101.2	66.6	18.4
North Central London	18,935	858	2,915	4,585	5,826	3,732	1,019	60.4	23.7	58.2	71.7	98.4	67.7	20.8
North East London	26,785	1,338	5,574	7,770	7,213	3,989	901	72.3	26.8	93.4	111.7	104.2	61.9	15.6
North West London	27,162	1,057	3,910	6,943	8,452	5,467	1,333	58.0	19.5	50.2	72.4	94.6	67.6	18.9
South East London	24,049	1,297	3,785	5,917	7,378	4,524	1,148	65.5	28.9	73.0	86.2	106.1	65.6	18.2
South West London	19,088	704	2,196	4,064	6,884	4,225	1,015	57.9	19.2	48.0	61.4	104.1	70.3	18.6
SOUTH EAST	93,921	5,132	14,578	22,709	30,118	17,654	3,730	57.3	20.3	61.6	96.1	109.4	55.6	11.6
Hampshire and Isle of Wight	19,833	1,326	3,413	5,073	6,071	3,256	694	54.5	23.2	59.1	97.8	102.6	48.0	9.9
Kent and Medway	18,747	1,325	3,599	4,872	5,397	2,961	593	58.5	25.3	79.6	109.2	101.9	47.4	9.4
Surrey and Sussex	28,181	1,245	3,578	6,249	9,681	6,105	1,323	56.2	16.4	52.7	88.6	115.2	60.9	12.9
Thames Valley	27,160	1,236	3,988	6,515	8,969	5,332	1,120	59.8	18.2	60.4	94.1	113.3	61.4	13.2
SOUTH WEST	52,729	3,306	9,284	12,917	16,223	9,070	1,929	55.0	21.1	66.3	99.6	103.8	48.8	10.1
Avon, Gloucestershire and Wiltshire	25,390	1,502	4,223	6,073	8,142	4,505	945	56.4	21.5	62.8	93.5	108.4	52.1	10.9
Dorset and Somerset	11,716	684	2,097	2,907	3,616	1,971	441	54.4	18.6	73.3	104.5	103.2	46.2	9.8
South West Peninsula	15,623	1,120	2,964	3,937	4,465	2,594	543	53.2	22.5	66.9	106.6	96.7	45.5	9.2

1 The rates for women of all ages, under 20 and 40 and over are based on women aged 15-44, 15-19 and 40-44 respectively.
2 See sections 2.1 and 2.11.

Table 7.4 - *continued*

Area of usual residence of mother	Age of mother at birth							Age of mother at birth						
	All ages	Under 20	20-24	25-29	30-34	35-39	40 and over	All ages[1]	Under 20	20-24	25-29	30-34	35-39	40 and over
	Numbers							Rates per 1,000 women in age group[2]						
WALES	32,593	3,099	7,228	8,393	8,553	4,416	904	**56.2**	32.0	77.1	105.0	92.8	41.3	8.2
Anglesey	685	63	151	197	166	86	22	**56.9**	30.5	92.8	114.6	84.7	37.7	9.2
Gwynedd	1,264	116	259	317	327	197	48	**57.5**	30.5	64.4	107.5	94.5	51.2	12.3
Conwy	1,044	95	198	257	289	174	31	**55.9**	30.4	82.7	106.0	97.8	47.0	7.6
Denbighshire	984	92	212	256	259	136	29	**57.4**	31.3	100.0	107.2	95.7	40.6	8.0
Flintshire	1,642	136	323	408	454	270	51	**55.6**	30.2	82.4	100.0	89.6	45.4	8.5
Wrexham	1,582	153	336	400	432	231	30	**61.1**	39.3	83.9	106.1	98.3	46.3	6.2
Powys	1,248	77	188	324	367	237	55	**57.0**	20.4	72.3	121.2	103.5	52.9	11.4
Ceredigion	600	46	109	146	174	94	31	**38.8**	16.1	26.3	81.8	90.9	42.5	12.1
Pembrokeshire	1,201	110	281	313	295	174	28	**58.5**	30.5	103.8	112.8	91.3	45.0	6.4
Carmarthenshire	1,744	139	352	481	468	248	56	**54.0**	25.1	77.0	109.7	92.5	40.3	8.5
Swansea	2,449	253	569	605	638	315	69	**54.4**	32.8	65.4	100.7	92.1	40.7	8.7
Neath Port Talbot	1,486	154	356	411	389	145	31	**57.7**	35.1	94.5	121.3	95.3	29.9	5.8
Bridgend	1,523	145	336	404	416	184	38	**59.5**	34.4	98.3	120.4	93.9	36.9	7.3
Vale of Glamorgan	1,279	84	244	324	356	222	49	**53.4**	20.3	76.1	101.2	94.4	47.6	9.9
Cardiff	3,955	339	850	917	1,113	607	129	**51.3**	28.9	46.8	73.6	97.3	51.7	11.2
Rhondda Cynon Taff	2,864	327	723	809	671	289	45	**60.7**	42.0	96.5	119.4	85.5	33.3	5.2
Merthyr Tydfil	643	95	217	146	127	46	12	**58.6**	49.5	127.1	109.9	73.7	21.7	5.5
Caerphilly	2,055	237	500	559	516	207	36	**60.1**	41.9	102.8	113.2	87.5	31.8	5.7
Blaenau Gwent	734	91	234	198	128	75	8	**54.1**	36.4	121.6	112.9	60.5	28.5	3.0
Torfaen	1,094	141	280	295	241	116	21	**62.4**	46.0	114.2	123.4	86.6	34.9	6.0
Monmouthshire	819	43	114	184	277	158	43	**53.4**	16.2	68.2	107.0	116.5	47.4	12.0
Newport	1,698	163	396	442	450	205	42	**60.0**	33.6	93.7	119.6	99.3	36.5	7.8
Normal residence outside England and Wales	214	8	35	58	59	34	20	:	:	:	:	:	:	:

1 The rates for women of all ages, under 20 and 40 and over are based on women aged 15-44, 15-19 and 40-44 respectively.
2 See sections 2.1 and 2.11.

Table 7.5   Live births (numbers and percentages):   **England and Wales, Government**
             birthweight and area of usual residence, 2005   **Office Regions (within England)**

Area of usual residence of mother	Birthweight (grams)												
	All weights[1]	Under 1,500	1,500-1,999	2,000-2,499	2,500-2,999	3,000-3,499	3,500 and over	All weights[1]	Under 1,500	1,500-1,999	2,000-2,499	2,500-2,999	3,000-3,499	3,500 and over
	Numbers							**Percentages**						
ENGLAND AND WALES	**645,835**	**8,045**	**9,955**	**30,791**	**109,454**	**230,010**	**255,339**	**100**	**1.2**	**1.5**	**4.8**	**16.9**	**35.6**	**39.5**
ENGLAND	**613,028**	**7,636**	**9,458**	**29,411**	**104,149**	**218,353**	**241,800**	**100**	**1.2**	**1.5**	**4.8**	**17.0**	**35.6**	**39.4**
North East	**28,249**	367	460	1,384	4,674	9,785	11,576	**100**	1.3	1.6	4.9	16.5	34.6	41.0
North West	**81,722**	1,073	1,353	3,925	14,073	28,776	32,425	**100**	1.3	1.7	4.8	17.2	35.2	39.7
Yorkshire and the Humber	**60,665**	803	954	3,027	10,915	21,543	23,366	**100**	1.3	1.6	5.0	18.0	35.5	38.5
East Midlands	**49,080**	580	775	2,353	8,260	17,392	19,698	**100**	1.2	1.6	4.8	16.8	35.4	40.1
West Midlands	**65,956**	908	1,115	3,548	12,177	23,642	24,521	**100**	1.4	1.7	5.4	18.5	35.8	37.2
East	**64,687**	686	881	2,783	10,434	22,782	27,044	**100**	1.1	1.4	4.3	16.1	35.2	41.8
London	**116,019**	1,589	1,781	6,191	21,020	42,693	40,970	**100**	1.4	1.5	5.3	18.1	36.8	35.3
South East	**93,921**	1,076	1,413	3,962	14,630	33,304	39,410	**100**	1.1	1.5	4.2	15.6	35.5	42.0
South West	**52,729**	554	726	2,238	7,966	18,436	22,790	**100**	1.1	1.4	4.2	15.1	35.0	43.2
WALES	**32,593**	383	483	1,363	5,274	11,583	13,490	**100**	1.2	1.5	4.2	16.2	35.5	41.4
Normal residence outside **England and Wales**	214	26	14	17	31	74	49	**100**	12.1	6.5	7.9	14.5	34.6	22.9

1 'All weights' includes births where the birthweight was not stated.

Table 7.6   **Stillbirths (numbers and rates): occurrence within/outside marriage, and area of usual residence, 2005**

Area of usual residence of mother	Stillbirths			Stillbirths per 1,000 live and stillbirths[1]
	Total	Within marriage	Outside marriage	
ENGLAND AND WALES	**3,483**	**1,861**	**1,622**	**5.4**
ENGLAND	**3,298**	**1,773**	**1,525**	**5.4**
North East	**162**	58	104	5.7
North West	**453**	206	247	5.5
Yorkshire and the Humber	**379**	202	177	6.2
East Midlands	**245**	118	127	5.0
West Midlands	**394**	218	176	5.9
East	**282**	168	114	4.3
London	**702**	426	276	6.0
South East	**454**	266	188	4.8
South West	**227**	111	116	4.3
WALES	**175**	81	94	5.3
Normal residence outside England and Wales	**10**	7	3	44.6

1 See section 2.11.

Table 7.7   **Stillbirths (numbers and percentages): birthweight and area of usual residence, 2005**

Area of usual residence of mother	Birthweight (grams)													
	All weights[1]	Under 1,500	1,500-1,999	2,000-2,499	2,500-2,999	3,000-3,499	3,500 and over	**All weights**[1]	Under 1,500	1,500-1,999	2,000-2,499	2,500-2,999	3,000-3,499	3,500 and over
	Numbers							**Percentages**						
ENGLAND AND WALES	**3,483**	**1,565**	**374**	**383**	**418**	**390**	**298**	**100**	**44.9**	**10.7**	**11.0**	**12.0**	**11.2**	**8.6**
ENGLAND	**3,298**	**1,478**	**354**	**365**	**399**	**367**	**280**	**100**	**44.8**	**10.7**	**11.1**	**12.1**	**11.1**	**8.5**
North East	**162**	72	15	20	20	17	15	**100**	44.4	_9.3_	12.3	12.3	_10.5_	_9.3_
North West	**453**	188	52	49	64	51	38	**100**	41.5	11.5	10.8	14.1	11.3	8.4
Yorkshire and the Humber	**379**	180	40	39	48	43	25	**100**	47.5	10.6	10.3	12.7	11.3	6.6
East Midlands	**245**	110	23	31	28	21	29	**100**	44.9	9.4	12.7	11.4	8.6	11.8
West Midlands	**394**	183	46	41	42	51	28	**100**	46.4	11.7	10.4	10.7	12.9	7.1
East	**282**	125	38	33	36	26	21	**100**	44.3	13.5	11.7	12.8	9.2	7.4
London	**702**	339	76	63	73	67	65	**100**	48.3	10.8	9.0	10.4	9.5	9.3
South East	**454**	180	39	64	61	62	43	**100**	39.6	8.6	14.1	13.4	13.7	9.5
South West	**227**	101	25	25	27	29	16	**100**	44.5	11.0	11.0	11.9	12.8	_7.0_
WALES	**175**	78	19	18	19	23	18	**100**	44.6	_10.9_	_10.3_	_10.9_	13.1	_10.3_
Normal residence outside England and Wales	**10**	9	1	-	-	-	-	**100**	_90.0_	_10.0_	-	-	-	-

1 'All weights' includes births where the birthweight was not stated.

Table 8.1   Maternities: age of mother, occurrence within/outside marriage, number of previous live-born children[1] and place of confinement, 2005   **England and Wales**

	Place of confinement[2]	Age of mother at birth							
		All ages	Under 20	20-24	25-29	30-34	35-39	40-44	45 and over
Total	**Total**	**639,627**	**44,813**	**121,644**	**162,986**	**185,749**	**102,536**	**20,855**	**1,044**
	NHS Hospitals	**619,285**	**44,340**	**119,271**	**158,250**	**178,723**	**97,811**	**19,889**	**1,001**
	Non-NHS Hospitals	**2,905**	**26**	**273**	**540**	**1,051**	**794**	**201**	**20**
	At Home	**16,501**	**363**	**1,868**	**3,942**	**5,771**	**3,804**	**730**	**23**
	Elsewhere	**936**	**84**	**232**	**254**	**204**	**127**	**35**	-
All born within marriage	**Total**	**365,089**	**3,656**	**39,862**	**99,120**	**135,571**	**72,622**	**13,552**	**706**
	NHS Hospitals	**351,749**	3,610	38,948	96,205	130,313	69,117	12,880	676
	Non-NHS Hospitals	**2,614**	16	242	482	973	715	168	18
	At Home	**10,337**	27	616	2,319	4,166	2,710	487	12
	Elsewhere	**389**	3	56	114	119	80	17	-
Previous live-born **0**	**Total**	**153,844**	**2,965**	**22,051**	**49,528**	**54,797**	**20,870**	**3,423**	**210**
	NHS Hospitals	**150,728**	2,939	21,733	48,704	53,543	20,291	3,320	198
	Non-NHS Hospitals	**1,225**	15	159	279	474	238	50	10
	At Home	**1,816**	10	139	523	754	337	51	2
	Elsewhere	**75**	1	20	22	26	4	2	-
1	**Total**	**130,654**	**623**	**13,163**	**31,817**	**52,175**	**28,165**	**4,547**	**164**
	NHS Hospitals	**124,982**	606	12,775	30,560	49,800	26,749	4,331	161
	Non-NHS Hospitals	**912**	1	74	143	355	281	56	2
	At Home	**4,576**	14	291	1,058	1,959	1,101	152	1
	Elsewhere	**184**	2	23	56	61	34	8	-
2	**Total**	**51,687**	**63**	**3,730**	**12,235**	**18,482**	**14,287**	**2,764**	**126**
	NHS Hospitals	**48,704**	61	3,568	11,667	17,377	13,323	2,584	124
	Non-NHS Hospitals	**330**	-	9	40	111	131	38	1
	At Home	**2,577**	2	145	507	971	813	138	1
	Elsewhere	**76**	-	8	21	23	20	4	-
3	**Total**	**17,926**	**5**	**771**	**3,998**	**6,314**	**5,361**	**1,396**	**81**
	NHS Hospitals	**16,875**	4	735	3,795	5,959	4,999	1,308	75
	Non-NHS Hospitals	**96**	-	-	18	20	40	15	3
	At Home	**920**	1	31	175	329	309	72	3
	Elsewhere	**35**	-	5	10	6	13	1	-
4	**Total**	**6,323**	-	**123**	**1,110**	**2,343**	**2,066**	**627**	**54**
	NHS Hospitals	**5,994**	-	114	1,063	2,229	1,957	582	49
	Non-NHS Hospitals	**30**	-	-	2	11	14	1	2
	At Home	**289**	-	9	40	103	90	44	3
	Elsewhere	**10**	-	-	5	-	5	-	-
4 and over	**Total**	**10,978**	-	**147**	**1,542**	**3,803**	**3,939**	**1,422**	**125**
	NHS Hospitals	**10,460**	-	137	1,479	3,634	3,755	1,337	118
	Non-NHS Hospitals	**51**	-	-	2	13	25	9	2
	At Home	**448**	-	10	56	153	150	74	5
	Elsewhere	**19**	-	-	5	3	9	2	-
5-9	**Total**	**4,513**	-	**22**	**430**	**1,449**	**1,809**	**738**	**65**
	NHS Hospitals	**4,335**	-	21	414	1,395	1,739	703	63
	Non-NHS Hospitals	**18**	-	-	-	2	10	6	-
	At Home	**151**	-	1	16	49	56	27	2
	Elsewhere	**9**	-	-	-	3	4	2	-
10-14	**Total**	**137**	-	**2**	**2**	**11**	**64**	**53**	**5**
	NHS Hospitals	**126**	-	2	2	10	59	48	5
	Non-NHS Hospitals	**3**	-	-	-	-	1	2	-
	At Home	**8**	-	-	-	1	4	3	-
	Elsewhere	-	-	-	-	-	-	-	-
15 and over	**Total**	**5**	-	-	-	-	-	**4**	**1**
	NHS Hospitals	**5**	-	-	-	-	-	4	1
	Non-NHS Hospitals	-	-	-	-	-	-	-	-
	At Home	-	-	-	-	-	-	-	-
	Elsewhere	-	-	-	-	-	-	-	-
All born outside marriage	**Total**	**274,538**	**41,157**	**81,782**	**63,866**	**50,178**	**29,914**	**7,303**	**338**
	NHS Hospitals	**267,536**	40,730	80,323	62,045	48,410	28,694	7,009	325
	Non-NHS Hospitals	**291**	10	31	58	78	79	33	2
	At Home	**6,164**	336	1,252	1,623	1,605	1,094	243	11
	Elsewhere	**547**	81	176	140	85	47	18	-

1   See section 2.9.
2   See section 3.7.

Table 8.2 Maternities: place of confinement and whether area of occurrence is the same as area of usual residence, or other than area of usual residence, 2005

England and Wales, Government Office Regions (within England), and health authorities/boards

Area of usual residence	Place of confinement[1]					Health authority/board of occurrence		
	Total	NHS hospitals	Non-NHS hospitals	At home	Elsewhere	Same as usual residence	Other than usual residence	At home and elsewhere
ENGLAND AND WALES	**639,627**	**619,285**	**2,905**	**16,501**	**936**	**361,441**	**278,186**	-
ENGLAND	**607,090**	**588,037**	**2,864**	**15,335**	**854**	**342,833**	**264,257**	-
NORTH EAST	**27,971**	**27,601**	**1**	**324**	**45**	**26,717**	**885**	**369**
County Durham and Tees Valley	**13,116**	12,955	-	139	22	12,194	761	161
Northumberland, Tyne & Wear	**14,855**	14,646	1	185	23	14,523	124	208
NORTH WEST	**80,984**	**79,462**	**3**	**1,452**	**67**	**75,993**	**3,472**	**1,519**
Cheshire & Merseyside	**26,069**	25,656	1	387	25	23,574	2,083	412
Cumbria and Lancashire	**21,627**	21,214	2	400	11	20,218	998	411
Greater Manchester	**33,288**	32,592	-	665	31	32,201	391	696
YORKSHIRE AND THE HUMBER	**60,198**	**58,994**	**5**	**1,102**	**97**	**56,979**	**2,020**	**1,199**
North and East Yorkshire and Northern Lincolnshire	**17,277**	16,922	3	321	31	15,875	1,050	352
South Yorkshire	**14,943**	14,627	1	290	25	14,256	372	315
West Yorkshire	**27,978**	27,445	1	491	41	26,848	598	532
EAST MIDLANDS	**48,603**	**47,238**	**13**	**1,276**	**76**	**41,765**	**5,486**	**1,352**
Leicestershire, Northamptonshire and Rutland	**19,694**	19,123	12	528	31	16,752	2,383	559
Trent	**28,909**	28,115	1	748	45	25,013	3,103	793
WEST MIDLANDS	**65,428**	**64,204**	**4**	**1,146**	**74**	**60,730**	**3,314**	**1,384**
Birmingham and the Black Country	**31,945**	31,509	-	409	27	31,271	238	436
Shropshire and Staffordshire	**16,496**	16,135	1	335	25	13,984	2,050	462
West Midlands South	**16,987**	16,560	3	402	22	15,475	1,026	486
EAST	**63,966**	**60,919**	**611**	**2,316**	**120**	**56,120**	**5,410**	**2,436**
Bedfordshire and Hertfordshire	**20,817**	20,114	58	616	29	16,819	3,353	645
Essex	**18,836**	18,070	38	690	38	16,233	1,875	728
Norfolk, Suffolk and Cambridgeshire	**24,313**	22,735	515	1,010	53	23,068	182	1,063
LONDON	**114,863**	**110,310**	**2,050**	**2,369**	**134**	**101,293**	**11,067**	**2,503**
North Central London	**18,707**	17,746	611	333	17	15,972	2,385	350
North East London	**26,592**	26,015	112	441	24	24,562	1,565	465
North West London	**26,898**	25,493	1,062	317	26	24,731	1,824	343
South East London	**23,785**	22,802	72	867	44	21,582	1,292	911
South West London	**18,881**	18,254	193	411	23	14,446	4,001	434
SOUTH EAST	**92,899**	**89,366**	**164**	**3,220**	**149**	**82,113**	**7,417**	**3,369**
Hampshire and Isle of Wight	**19,651**	18,891	16	712	32	15,621	3,286	744
Kent and Medway	**18,545**	17,710	32	770	33	17,061	681	803
Surrey and Sussex	**27,825**	26,755	67	946	57	25,246	1,576	1,003
Thames Valley	**26,878**	26,010	49	792	27	24,185	1,874	819
SOUTH WEST	**52,178**	**49,943**	**13**	**2,130**	**92**	**48,270**	**1,686**	**2,222**
Avon, Gloucestershire and Wiltshire	**25,132**	24,343	8	750	31	23,961	390	781
Dorset and Somerset	**11,583**	11,084	1	477	21	9,947	1,138	498
South West Peninsula	**15,463**	14,516	4	903	40	14,362	158	943
WALES	**32,325**	**31,083**	**1**	**1,166**	**75**	**18,601**	**12,483**	**1,241**
Anglesey	**680**	659	-	21	-	-	659	21
Gwynedd	**1,250**	1,183	-	60	7	1,038	145	67
Conwy	**1,032**	1,020	-	11	1	-	1,020	12
Denbighshire	**979**	941	-	36	2	866	75	38
Flintshire	**1,630**	1,607	-	21	2	-	1,607	23

1 See section 3.7.

Table 8.2 - *continued*

Area of usual residence	Place of confinement[1]					Health authority/board of occurrence		
	Total	NHS hospitals	Non-NHS hospitals	At home	Elsewhere	Same as usual residence	Other than usual residence	At home and elsewhere
WALES - *continued*								
Wrexham	**1,571**	1,541	-	28	2	1,483	58	30
Powys	**1,234**	1,118	-	109	7	164	954	116
Ceredigion	**594**	563	-	31	-	390	173	31
Pembrokeshire	**1,189**	1,113	-	72	4	1,064	49	76
Carmarthenshire	**1,734**	1,634	-	97	3	1,179	455	100
Swansea	**2,426**	2,341	-	80	5	2,290	51	85
Neath Port Talbot	**1,470**	1,412	-	54	4	296	1,116	58
Bridgend	**1,513**	1,370	-	136	7	1,287	83	143
Vale of Glamorgan	**1,266**	1,215	-	47	4	594	621	51
Cardiff	**3,927**	3,825	-	97	5	2,939	886	102
Rhondda Cynon Taff	**2,842**	2,761	-	75	6	2,208	553	81
Merthyr Tydfil	**642**	636	-	6	-	591	45	6
Caerphilly	**2,047**	2,001	-	40	6	341	1,660	46
Blaenau Gwent	**732**	712	1	18	1	-	713	19
Torfaen	**1,082**	1,049	-	28	5	-	1,049	33
Monmouthshire	**807**	759	-	47	1	350	409	48
Newport	**1,678**	1,623	-	52	3	1,521	102	55
Normal residence outside England and Wales	**212**	165	40	-	7	7	205	-

1 See section 3.7.

Table 8.3   Maternities in hospitals: live births and stillbirths, and area of occurrence[1], 2005

England and Wales, Government Office Regions (within England), and health authorities/boards

Area of occurrence	Maternities	Live births	Stillbirths	Stillbirths per 1,000 live births and stillbirths[2]
ENGLAND AND WALES	**622,190**	**628,419**	**3,428**	**5.4**
ENGLAND	**592,099**	**598,059**	**3,268**	**5.4**
NORTH EAST	**27,739**	**28,027**	**154**	**5.5**
County Durham and Tees Valley	12,472	12,595	76	6.0
Northumberland, Tyne & Wear	15,267	15,432	78	5.0
NORTH WEST	**80,954**	**81,714**	**464**	**5.6**
Cheshire & Merseyside	24,980	25,290	127	5.0
Cumbria and Lancashire	21,962	22,120	119	5.4
Greater Manchester	34,012	34,304	218	6.3
YORKSHIRE AND THE HUMBER	**59,516**	**59,997**	**378**	**6.3**
North and East Yorkshire and Northern Lincolnshire	16,794	16,955	77	4.5
South Yorkshire	14,815	14,962	91	6.0
West Yorkshire	27,907	28,080	210	7.4
EAST MIDLANDS	**43,022**	**43,453**	**216**	**4.9**
Leicestershire, Northamptonshire and Rutland	16,966	17,141	92	5.3
Trent	26,056	26,312	124	4.7
WEST MIDLANDS	**65,660**	**66,197**	**398**	**6.0**
Birmingham and the Black Country	33,497	33,761	219	6.4
Shropshire and Staffordshire	15,628	15,755	92	5.8
West Midlands South	16,535	16,681	87	5.2
EAST	**60,043**	**60,736**	**275**	**4.5**
Bedfordshire and Hertfordshire	17,581	17,777	96	5.4
Essex	17,086	17,288	87	5.0
Norfolk, Suffolk and Cambridgeshire	25,376	25,671	92	3.6
LONDON	**115,348**	**116,559**	**706**	**6.0**
North Central London	19,031	19,242	104	5.4
North East London	26,326	26,524	191	7.1
North West London	30,257	30,595	179	5.8
South East London	23,574	23,855	141	5.9
South West London	16,160	16,343	91	5.5
SOUTH EAST	**89,045**	**90,047**	**446**	**4.9**
Hampshire and Isle of Wight	16,110	16,272	92	5.6
Kent and Medway	17,615	17,812	86	4.8
Surrey and Sussex	29,579	29,933	129	4.3
Thames Valley	25,741	26,030	139	5.3
SOUTH WEST	**50,772**	**51,329**	**231**	**4.5**
Avon, Gloucestershire and Wiltshire	25,995	26,274	131	5.0
Dorset and Somerset	10,269	10,389	34	3.3
South West Peninsula	14,508	14,666	66	4.5
WALES	**30,091**	**30,360**	**160**	**5.2**
Anglesey	-	-	-	-
Gwynedd	1,900	1,922	9	4.7
Conwy	-	-	-	-
Denbighshire	2,247	2,267	8	3.5
Flintshire	-	-	-	-
Wrexham	2,321	2,337	15	6.4
Powys	171	171	-	-
Ceredigion	526	532	2	3.7
Pembrokeshire	1,126	1,137	3	2.6
Carmarthenshire	1,345	1,352	10	7.3

1 See section 3.7.
2 See section 2.11.

Table 8.3 - *continued*

Area of occurrence	Maternities	Live births	Stillbirths	Stillbirths per 1,000 live births and stillbirths[2]
WALES - *continued*				
Swansea	3,465	3,511	20	5.7
Neath Port Talbot	330	330	-	-
Bridgend	1,946	1,959	11	5.6
Vale of Glamorgan	1,348	1,348	11	8.1
Cardiff	3,879	3,916	22	5.6
Rhondda Cynon Taff	2,480	2,506	11	4.4
Merthyr Tydfil	1,284	1,296	7	5.4
Caerphilly	388	388	-	-
Blaenau Gwent	-	-	-	-
Torfaen	-	-	-	-
Monmouthshire	1,886	1,897	11	5.8
Newport	3,449	3,491	20	5.7

2 See section 2.11.

Table 9.1 Live births (numbers and percentages):
country of birth of mother[2], 1995, 2000-2005

England and Wales

Country of birth of mother	1995	2000	2001	2002	2003	2004	2005
	Numbers						
Total	**648,138**	**604,441**	**594,634**	**596,122**	**621,469**	**639,721**	**645,835**
United Kingdom[1]	566,452	510,835	496,713	490,711	506,076	515,144	511,624
Total outside United Kingdom	**81,677**	**93,588**	**97,895**	**105,381**	**115,360**	**124,563**	**134,189**
Irish Republic	5,167	4,050	3,843	3,708	3,734	3,597	3,463
Australia, Canada and New Zealand	3,051	3,635	3,695	3,885	4,132	4,152	4,217
New Commonwealth	**44,103**	**45,348**	**47,963**	**51,500**	**54,983**	**59,199**	**62,404**
India	6,684	6,650	6,598	7,222	7,995	9,146	10,079
Pakistan	12,324	13,561	14,588	15,357	15,107	15,736	16,477
Bangladesh	6,783	7,482	8,164	8,486	8,898	8,857	8,217
East Africa	5,128	3,959	3,745	3,724	3,985	3,991	4,042
Southern Africa	1,010	1,907	2,236	2,654	3,149	3,722	4,122
Rest of Africa	5,934	5,646	5,857	6,445	7,456	9,264	11,029
Caribbean	2,912	2,681	3,085	3,598	3,984	3,816	3,733
Far East	1,996	1,538	1,365	1,351	1,433	1,439	1,341
Rest of New Commonwealth	1,332	1,924	2,325	2,663	2,976	3,228	3,364
Other European Union	11,113	13,829	13,840	14,576	15,830	17,290	20,420
Rest of Europe	2,360	5,468	5,467	5,662	6,249	7,072	7,504
United States of America	2,641	2,895	2,878	2,828	3,037	2,916	2,930
Rest of the World	13,242	18,363	20,209	23,222	27,395	30,337	33,251
Not stated	9	18	26	30	33	14	22
	Percentage of all live births						
Total	**100.0**	**100.0**	**100.0**	**100.0**	**100.0**	**100.0**	**100.0**
United Kingdom[1]	87.4	84.5	83.5	82.3	81.4	80.5	79.2
Total outside United Kingdom	**12.6**	**15.5**	**16.5**	**17.7**	**18.6**	**19.5**	**20.8**
Irish Republic	0.8	0.7	0.6	0.6	0.6	0.6	0.5
Australia, Canada and New Zealand	0.5	0.6	0.6	0.7	0.7	0.6	0.7
New Commonwealth	**6.8**	**7.5**	**8.1**	**8.6**	**8.8**	**9.3**	**9.7**
India	1.0	1.1	1.1	1.2	1.3	1.4	1.6
Pakistan	1.9	2.2	2.5	2.6	2.4	2.5	2.6
Bangladesh	1.0	1.2	1.4	1.4	1.4	1.4	1.3
East Africa	0.8	0.7	0.6	0.6	0.6	0.6	0.6
Southern Africa	0.2	0.3	0.4	0.4	0.5	0.6	0.6
Rest of Africa	0.9	0.9	1.0	1.1	1.2	1.4	1.7
Caribbean	0.4	0.4	0.5	0.6	0.6	0.6	0.6
Far East	0.3	0.3	0.2	0.2	0.2	0.2	0.2
Rest of New Commonwealth	0.2	0.3	0.4	0.4	0.5	0.5	0.5
Other European Union	1.7	2.3	2.3	2.4	2.5	2.7	3.2
Rest of Europe	0.4	0.9	0.9	0.9	1.0	1.1	1.2
United States of America	0.4	0.5	0.5	0.5	0.5	0.5	0.5
Rest of the World	2.0	3.0	3.4	3.9	4.4	4.7	5.1
Not stated	0.0	0.0	0.0	0.0	0.0	0.0	0.0

Note: For comparability, the births data for all years for mothers born outside the United Kingdom were reclassified according to the 2005 country
classification list and the definition of the European Union (EU25), as constituted in 2005, was used for all years' data.
1 Including Isle of Man and Channel Islands.
2 See Section 2.7.

Table 9.2 Live births (numbers and percentages): birthplace of mother if outside United Kingdom[1], and area of usual residence, 2005

England and Wales, Government Office Regions (within England), counties[2], unitary authorities, county districts and London boroughs, with more than 15 per cent non-UK born mothers

Area of usual residence of mother	All live births	Birthplace of mother outside United Kingdom					
		New Commonwealth		Rest of the World		All outside United Kingdom	
		Number	Percentage	Number	Percentage	Number	Percentage
ENGLAND AND WALES[3]	**645,621**	**62,383**	**10**	**71,726**	**11**	**134,109**	**21**
ENGLAND	**613,028**	**61,487**	**10**	**70,088**	**11**	**131,575**	**21**
NORTH EAST	**28,249**	**909**	**3**	**1,265**	**4**	**2,174**	**8**
Tyne and Wear (Met County)	**11,968**	538	4	686	6	1,224	10
Newcastle upon Tyne	**2,979**	282	9	337	11	619	21
NORTH WEST	**81,722**	**6,007**	**7**	**4,681**	**6**	**10,688**	**13**
Blackburn with Darwen UA	**2,284**	457	20	68	3	525	23
Greater Manchester (Met County)	**33,544**	3,852	11	2,598	8	6,450	19
Bolton	**3,576**	451	13	213	6	664	19
Manchester	**6,707**	1,302	19	1,107	17	2,409	36
Oldham	**3,200**	736	23	109	3	845	26
Rochdale	**2,853**	481	17	114	4	595	21
Lancashire	**13,084**	1,011	8	494	4	1,505	12
Burnley	**1,163**	161	14	46	4	207	18
Pendle	**1,244**	256	21	31	2	287	23
Preston	**1,801**	267	15	86	5	353	20
YORKSHIRE AND THE HUMBER	**60,665**	**5,249**	**9**	**3,618**	**6**	**8,867**	**15**
South Yorkshire (Met County)	**15,081**	833	6	1,016	7	1,849	12
Sheffield	**6,101**	571	9	620	10	1,191	20
West Yorkshire (Met County)	**28,142**	4,087	15	1,683	6	5,770	21
Bradford	**8,014**	2,074	26	445	6	2,519	31
Kirklees	**5,309**	845	16	216	4	1,061	20
Leeds	**8,709**	793	9	784	9	1,577	18
EAST MIDLANDS	**49,080**	**3,084**	**6**	**3,166**	**6**	**6,250**	**13**
Derby UA	**2,989**	332	11	207	7	539	18
Leicester UA	**4,597**	1,216	26	631	14	1,847	40
Nottingham UA	**3,753**	444	12	344	9	788	21
Leicestershire	**6,657**	290	4	303	5	593	9
Oadby and Wigston	**573**	67	12	25	4	92	16
Northamptonshire	**8,279**	438	5	700	8	1,138	14
Northampton	**2,828**	267	9	350	12	617	22
WEST MIDLANDS	**65,956**	**7,708**	**12**	**4,431**	**7**	**12,139**	**18**
Stoke-on-Trent UA	**3,312**	328	10	191	6	519	16
West Midlands (Met County)	**36,038**	6,536	18	2,997	8	9,533	26
Birmingham	**15,893**	4,024	25	1,685	11	5,709	36
Coventry	**3,871**	570	15	457	12	1,027	27
Sandwell	**4,171**	747	18	259	6	1,006	24
Walsall	**3,424**	500	15	116	3	616	18
Wolverhampton	**3,105**	360	12	202	7	562	18
EAST	**64,687**	**4,297**	**7**	**6,208**	**10**	**10,505**	**16**
Luton UA	**3,196**	1,070	33	398	12	1,468	46
Peterborough UA	**2,446**	311	13	365	15	676	28
Thurrock UA	**2,211**	219	10	139	6	358	16
Bedfordshire	**4,881**	377	8	388	8	765	16
Bedford	**1,922**	283	15	202	11	485	25
Cambridgeshire	**6,505**	274	4	937	14	1,211	19
Cambridge	**1,189**	100	8	323	27	423	36
East Cambridgeshire	**964**	23	2	152	16	175	18
South Cambridgeshire	**1,580**	68	4	203	13	271	17
Essex	**14,905**	506	3	960	6	1,466	10
Harlow	**1,109**	73	7	112	10	185	17

1 See section 2.7.
2 Counties which include county districts where the proportion of non-UK born mothers is more than 15 per cent have been included to allow for comparison.
3 This table excludes births to mothers whose usual residence was outside of England and Wales.

Table 9.2 - *continued*

Area of usual residence of mother	All live births	Birthplace of mother outside United Kingdom					
		New Commonwealth		Rest of the World		All outside United Kingdom	
		Number	Percentage	Number	Percentage	Number	Percentage
Hertfordshire	**12,975**	1,011	8	1,330	10	2,341	18
Dacorum	**1,623**	107	7	149	9	256	16
Hertsmere	**1,211**	122	10	149	12	271	22
St Albans	**1,769**	162	9	207	12	369	21
Three Rivers	**900**	66	7	96	11	162	18
Watford	**1,148**	209	18	173	15	382	33
Welwyn Hatfield	**1,223**	96	8	147	12	243	20
Suffolk	**7,493**	237	3	802	11	1,039	14
Forest Heath	**756**	11	1	343	45	354	47
Ipswich	**1,611**	119	7	140	9	259	16
LONDON	**116,019**	**26,484**	**23**	**32,968**	**28**	**59,452**	**51**
Inner London	**49,548**	12,247	25	16,544	33	28,791	58
Camden	**2,954**	505	17	1,303	44	1,808	61
City of London	**64**	9	14	24	38	33	52
Hackney	**4,375**	995	23	1,375	31	2,370	54
Hammersmith and Fulham	**2,686**	217	8	1,162	43	1,379	51
Haringey	**4,026**	700	17	1,778	44	2,478	62
Islington	**2,731**	318	12	999	37	1,317	48
Kensington and Chelsea	**2,188**	138	6	1,312	60	1,450	66
Lambeth	**4,739**	1,073	23	1,529	32	2,602	55
Lewisham	**4,284**	1,061	25	991	23	2,052	48
Newham	**5,353**	2,588	48	1,265	24	3,853	72
Southwark	**4,714**	1,530	32	1,249	26	2,779	59
Tower Hamlets	**3,968**	1,984	50	729	18	2,713	68
Wandsworth	**4,554**	710	16	1,228	27	1,938	43
Westminster, City of	**2,912**	419	14	1,600	55	2,019	69
Outer London	**66,471**	14,237	21	16,424	25	30,661	46
Barking and Dagenham	**2,985**	769	26	578	19	1,347	45
Barnet	**4,728**	715	15	1,671	35	2,386	50
Bexley	**2,686**	358	13	220	8	578	22
Brent	**4,503**	1,323	29	1,769	39	3,092	69
Bromley	**3,663**	287	8	460	13	747	20
Croydon	**4,704**	1,241	26	821	17	2,062	44
Ealing	**4,838**	1,145	24	1,778	37	2,923	60
Enfield	**4,496**	838	19	1,520	34	2,358	52
Greenwich	**3,963**	982	25	919	23	1,901	48
Harrow	**2,872**	979	34	711	25	1,690	59
Hillingdon	**3,489**	711	20	743	21	1,454	42
Hounslow	**3,674**	894	24	1,122	31	2,016	55
Kingston upon Thames	**2,000**	266	13	455	23	721	36
Merton	**2,925**	753	26	670	23	1,423	49
Redbridge	**3,577**	1,212	34	660	18	1,872	52
Richmond upon Thames	**2,580**	221	9	656	25	877	34
Sutton	**2,325**	275	12	357	15	632	27
Waltham Forest	**3,989**	1,097	28	1,111	28	2,208	55
SOUTH EAST	**93,921**	**6,165**	**7**	**9,895**	**11**	**16,060**	**17**
Bracknell Forest UA	**1,431**	82	6	188	13	270	19
Brighton and Hove UA	**3,045**	120	4	449	15	569	19
Milton Keynes UA	**3,200**	478	15	353	11	831	26
Reading UA	**2,146**	347	16	341	16	688	32
Slough UA	**2,108**	636	30	445	21	1,081	51
Southampton UA	**2,775**	195	7	338	12	533	19
Windsor and Maidenhead UA	**1,699**	143	8	276	16	419	25
Wokingham UA	**1,694**	146	9	193	11	339	20
Buckinghamshire	**5,511**	588	11	521	9	1,109	20
Aylesbury Vale	**1,956**	147	8	172	9	319	16
Chiltern	**832**	72	9	77	9	149	18
South Bucks	**615**	47	8	83	13	130	21
Wycombe	**2,108**	322	15	189	9	511	24
Hampshire	**13,537**	468	3	1,114	8	1,582	12
Rushmoor	**1,238**	83	7	162	13	245	20
Kent	**15,613**	616	4	1,226	8	1,842	12
Gravesham	**1,203**	121	10	76	6	197	16
Oxfordshire	**7,575**	485	6	1,033	14	1,518	20
Cherwell	**1,807**	92	5	199	11	291	16
Oxford	**1,782**	257	14	431	24	688	39
Vale of White Horse	**1,359**	61	4	165	12	226	17

Table 9.2 - *continued*

Area of usual residence of mother	All live births	Birthplace of mother outside United Kingdom					
		New Commonwealth		Rest of the World		All outside United Kingdom	
		Number	Percentage	Number	Percentage	Number	Percentage
Surrey	**12,303**	860	7	1,623	13	2,483	20
Elmbridge	**1,629**	133	8	286	18	419	26
Epsom and Ewell	**782**	67	9	99	13	166	21
Guildford	**1,436**	63	4	216	15	279	19
Reigate and Banstead	**1,491**	101	7	155	10	256	17
Runnymede	**840**	52	6	118	14	170	20
Spelthorne	**1,079**	92	9	144	13	236	22
Surrey Heath	**920**	63	7	147	16	210	23
Waverley	**1,223**	49	4	144	12	193	16
Woking	**1,252**	186	15	155	12	341	27
West Sussex	**8,050**	451	6	757	9	1,208	15
Crawley	**1,305**	215	16	141	11	356	27
Worthing	**1,115**	56	5	122	11	178	16
SOUTH WEST	**52,729**	**1,584**	**3**	**3,856**	**7**	**5,440**	**10**
Bournemouth UA	**1,619**	65	4	231	14	296	18
Bristol, City of UA	**5,440**	394	7	734	13	1,128	21
Swindon UA	**2,480**	185	7	228	9	413	17
WALES	**32,593**	896	3	1,638	5	2,534	8
Cardiff	**3,955**	345	9	443	11	788	20

Table 9.3   Live births: country of birth of mother and of father[2], 2005   **England and Wales**

Country of birth of father	Country of birth of mother									
	Total	United Kingdom[1]	**Total outside United Kingdom**	Irish Republic	Australia Canada and New Zealand	New Commonwealth				
						Total	India	Pakistan	Bangladesh	East Africa
Total	**645,835**	**511,624**	**134,189**	**3,463**	**4,217**	**62,404**	**10,079**	**16,477**	**8,217**	**4,042**
United Kingdom[1]	**473,906**	434,419	**39,480**	2,295	2,744	**15,990**	2,790	6,096	705	1,254
Total outside United Kingdom	**126,752**	**38,739**	**88,012**	**962**	**1,413**	**43,643**	**7,246**	**10,254**	**7,452**	**2,419**
Irish Republic	**2,990**	2,063	**927**	587	43	**66**	5	4	-	16
Australia, Canada and New Zealand	**3,973**	2,369	**1,604**	49	947	**123**	4	3	1	12
New Commonwealth	**65,268**	**20,692**	**44,575**	**128**	**150**	**41,590**	**7,114**	**10,052**	**7,434**	**2,038**
India	**9,170**	2,171	**6,999**	7	12	**6,726**	6,417	41	11	155
Pakistan	**18,029**	7,613	**10,416**	14	3	**10,054**	60	9,870	18	70
Bangladesh	**8,964**	1,481	**7,483**	5	1	**7,426**	20	5	7,389	5
East Africa	**4,259**	1,462	**2,797**	15	23	**2,282**	436	112	7	1,609
Southern Africa	**3,674**	1,465	**2,209**	20	44	**1,750**	20	7	0	34
Rest of Africa	**11,254**	2,435	**8,818**	27	18	**8,060**	53	8	3	115
Caribbean	**5,154**	3,101	**2,053**	22	13	**1,770**	4	2	2	38
Far East	**1,234**	578	**656**	15	18	**480**	25	2	0	5
Rest of New Commonwealth	**3,530**	386	**3,144**	3	18	**3,042**	79	5	4	7
Other European Union	**14,751**	6,392	**8,359**	83	114	**518**	32	45	5	55
Rest of Europe	**6,885**	1,542	**5,343**	17	26	**69**	1	1	2	12
United States of America	**2,641**	1,110	**1,531**	29	53	**74**	4	3	-	13
Rest of the World	**30,244**	4,571	**25,673**	69	80	**1,203**	86	146	10	273
Not stated	**45,177**	38,466	**6,697**	206	60	**2,771**	43	127	60	369

Country of birth of father	Country of birth of mother								
	New Commonwealth - *continued*					Other European Union	Rest of Europe	United States of America	Rest of the World	Not stated
	Southern Africa	Rest of Africa	Caribbean	Far East	Rest of New Common- wealth					
Total	**4,122**	**11,029**	**3,733**	**1,341**	**3,364**	**20,420**	**7,504**	**2,930**	**33,251**	**22**
United Kingdom[1]	1,632	1,504	1,077	613	319	9,539	1,727	1,445	5,740	7
Total outside United Kingdom	**2,273**	**8,297**	**1,994**	**701**	**3,007**	**9,868**	**5,476**	**1,427**	**25,223**	**1**
Irish Republic	14	9	8	10	-	103	19	22	87	-
Australia, Canada and New Zealand	61	7	13	14	8	247	54	75	109	-
New Commonwealth	**1,822**	**7,847**	**1,832**	**495**	**2,956**	**1,151**	**165**	**89**	**1,302**	**1**
India	16	35	9	20	22	114	21	10	109	-
Pakistan	9	14	4	6	3	151	31	8	155	-
Bangladesh	3	2	1	-	1	19	2	1	29	-
East Africa	55	35	14	4	10	129	22	14	312	-
Southern Africa	1,648	30	7	3	1	175	24	18	178	-
Rest of Africa	70	7,661	139	10	1	358	37	16	302	1
Caribbean	8	59	1,649	3	5	131	11	16	90	-
Far East	9	2	1	432	4	29	9	4	101	-
Rest of New Commonwealth	4	9	8	17	2,909	45	8	2	26	-
Other European Union	114	179	43	27	18	6,405	322	130	787	-
Rest of Europe	21	16	6	9	1	452	4,613	28	138	-
United States of America	18	9	14	9	4	180	55	987	153	-
Rest of the World	223	230	78	137	20	1,330	248	96	22,647	-
Not stated	217	1,228	662	27	38	1,013	301	58	2,288	14

1 Including Isle of Man and Channel Islands.
2 See section 2.7.

Table 9.4   Live births: age of mother and country of birth of mother[2], 2005   **England and Wales**

Country of birth of mother	Age of mother at birth							
	All ages	Under 20	20-24	25-29	30-34	35-39	40-44	45 and over
Total	**645,835**	**44,830**	**122,145**	**164,348**	**188,153**	**104,113**	**21,155**	**1,091**
United Kingdom[1]	**511,624**	40,851	98,691	123,591	147,539	83,447	16,760	745
Total outside United Kingdom	**134,189**	**3,979**	**23,447**	**40,751**	**40,609**	**20,663**	**4,394**	**346**
Irish Republic	**3,463**	98	366	646	1,189	969	186	9
Australia Canada and New Zealand	**4,217**	34	184	703	1,911	1,127	240	18
New Commonwealth	**62,404**	**1,564**	**12,231**	**21,270**	**17,633**	**7,810**	**1,726**	**170**
India	**10,079**	121	1,693	4,031	3,055	948	216	15
Pakistan	**16,477**	471	4,410	5,944	3,816	1,543	261	32
Bangladesh	**8,217**	199	2,589	2,954	1,802	562	100	11
East Africa	**4,042**	109	417	1,011	1,365	936	191	13
Southern Africa	**4,122**	79	490	1,183	1,654	610	102	4
Rest of Africa	**11,029**	272	1,241	3,736	3,449	1,786	491	54
Caribbean	**3,733**	253	761	939	974	578	199	29
Far East	**1,341**	9	72	296	483	388	86	7
Rest of New Commonwealth	**3,364**	51	558	1,176	1,035	459	80	5
Other European Union	**20,420**	704	3,347	5,710	6,487	3,513	624	35
Rest of Europe	**7,504**	279	1,557	2,426	2,096	960	181	5
United States of America	**2,930**	44	381	609	883	782	220	11
Rest of the World	**33,251**	1,256	5,381	9,387	10,410	5,502	1,217	98
Not stated	**22**	-	7	6	5	3	1	-

1 Including Isle of Man and Channel Islands.
2 See section 2.7.

Table 9.5   Total fertility rates: country of birth of mother, 1991 and 2001   **England and Wales**

Country of birth of mother	1991	2001
Total	**1.8**	**1.6**
United Kingdom[1]	1.8	1.6
Total outside United Kingdom	**2.3**	**2.2**
New Commonwealth	**2.8**	**2.8**
India	2.5	2.3
Pakistan	4.8	4.7
Bangladesh	5.3	3.9
East Africa	1.9	1.6
Rest of Africa[2]	2.7	2.0
Rest of New Commonwealth[3]	1.9	2.2
Rest of the World	1.9	1.8

Note: See sections 1.1, 2.7 and 2.11.
1 Including Isle of Man and Channel Islands.
2 Includes countries listed under Southern Africa and Rest of Africa in section 2.7.
3 Includes countries listed under Far East, Caribbean and Rest of New Commonwealth in section 2.7.

Table 9.6 Live births (numbers and percentages): occurrence within/outside marriage, number of previous live-born children and country of birth of mother[3], 1995, 2002-2005

England and Wales

Country of birth of mother	Year	All live births	All live births within marriage	Percentages							Births outside marriage
				Number of previous live-born children within marriage							
				0	1	2	3	4	5 and over		
Total[1]	1995	648,138	66.1	25.9	24.4	10.3	3.4	1.1	0.9		33.9
	2002	596,122	59.4	24.4	21.9	8.4	2.9	1.0	0.8		40.6
	2003	621,469	58.6	24.3	21.4	8.4	2.8	1.0	0.7		41.4
	2004	639,721	57.8	24.2	20.9	8.2	2.9	1.0	0.7		42.2
	2005	645,835	57.2	24.1	20.4	8.1	2.8	1.0	0.7		42.8
United Kingdom[2]	1995	566,452	63.4	25.1	24.0	9.8	3.1	0.9	0.5		36.6
	2002	490,711	54.6	22.2	21.1	7.6	2.4	0.8	0.5		45.4
	2003	506,076	53.6	22.0	20.5	7.5	2.3	0.7	0.5		46.4
	2004	515,144	52.5	21.6	19.9	7.4	2.3	0.7	0.5		47.5
	2005	511,624	51.5	21.3	19.4	7.3	2.3	0.7	0.5		48.5
Irish Republic	1995	5,167	69.8	28.8	23.1	10.9	3.6	1.4	2.0		30.2
	2002	3,708	67.4	28.2	23.3	9.5	3.1	1.4	2.0		32.6
	2003	3,734	67.9	29.1	22.5	10.1	3.2	1.3	1.7		32.1
	2004	3,597	66.2	27.8	23.8	8.8	3.2	1.2	1.4		33.8
	2005	3,463	67.7	29.3	23.0	9.4	3.3	1.2	1.4		32.3
New Commonwealth	1995	44,103	88.1	29.2	27.1	15.8	7.9	4.0	4.1		11.9
	2002	54,037	86.1	34.0	26.2	14.0	6.8	2.8	2.2		13.9
	2003	55,777	85.6	33.3	26.5	14.4	6.8	2.8	1.9		14.4
	2004	59,199	85.4	34.0	26.3	13.9	6.8	2.6	1.8		14.6
	2005	62,404	84.8	35.0	25.6	13.7	6.3	2.6	1.6		15.2
India	1995	6,684	97.3	35.6	34.1	17.3	6.4	2.5	1.4		2.7
	2002	7,222	97.8	47.9	33.1	11.3	3.6	1.2	0.7		2.2
	2003	7,995	97.6	49.2	32.8	11.0	3.2	0.9	0.6		2.4
	2004	9,146	98.2	51.6	33.1	9.8	2.7	0.7	0.4		1.8
	2005	10,079	98.1	53.5	31.9	9.1	2.5	0.8	0.3		1.9
Pakistan	1995	12,324	98.6	26.5	25.2	19.4	12.8	7.0	7.6		1.4
	2002	15,357	98.4	33.5	26.6	18.4	11.2	4.9	3.8		1.6
	2003	15,107	98.2	29.6	28.0	20.2	11.7	5.4	3.3		1.8
	2004	15,736	97.9	29.1	28.2	20.0	12.0	5.0	3.5		2.1
	2005	16,477	98.0	32.6	26.0	20.0	11.2	5.0	3.2		2.0
Bangladesh	1995	6,783	99.3	33.5	24.9	15.3	9.8	6.7	9.1		0.7
	2002	8,486	98.7	31.3	27.6	19.0	10.9	5.2	4.7		1.3
	2003	8,898	98.6	30.3	27.4	20.2	11.2	5.1	4.3		1.4
	2004	8,857	98.2	29.5	26.8	20.0	12.2	5.3	4.4		1.8
	2005	8,217	98.0	25.9	28.5	21.4	12.4	5.7	4.0		2.0
East Africa	1995	5,128	88.8	30.6	35.4	15.8	4.9	1.6	0.6		11.2
	2002	3,724	77.8	30.5	28.3	13.2	3.9	1.1	0.7		22.2
	2003	3,985	75.3	31.5	26.8	11.8	3.6	1.2	0.5		24.7
	2004	3,991	73.5	30.9	27.0	10.4	3.4	1.1	0.7		26.5
	2005	4,042	71.7	30.9	24.9	10.5	3.4	1.4	0.6		28.3
Southern Africa	1995	1,010	76.5	35.4	27.3	10.5	2.4	0.6	0.3		23.5
	2002	2,654	79.7	44.8	26.0	6.3	2.0	0.3	0.4		20.3
	2003	3,149	78.3	44.6	25.4	6.4	1.3	0.3	0.3		21.7
	2004	3,722	78.0	42.9	25.6	7.5	1.5	0.2	0.2		22.0
	2005	4,122	75.6	42.4	24.4	6.7	1.6	0.4	0.2		24.4
Rest of Africa	1995	5,934	61.4	20.6	20.3	12.7	5.3	1.8	0.7		38.6
	2002	8,073	64.9	26.5	20.4	11.6	4.3	1.4	0.8		35.1
	2003	7,456	64.1	26.1	19.5	12.0	4.8	1.1	0.6		35.9
	2004	9,264	63.3	27.8	18.7	11.0	4.3	1.2	0.4		36.7
	2005	11,029	63.8	28.4	19.0	10.8	4.0	1.1	0.4		36.2
Caribbean	1995	2,912	53.4	17.8	18.1	10.4	3.8	1.8	1.5		46.6
	2002	3,598	38.0	15.3	12.2	6.2	2.4	1.1	0.8		62.0
	2003	3,984	37.4	15.7	12.7	5.3	2.5	0.8	0.5		62.6
	2004	3,816	40.6	18.0	13.2	5.4	2.5	0.8	0.7		59.4
	2005	3,733	40.9	18.3	13.3	5.8	2.2	0.8	0.5		59.1
Far East	1995	1,996	86.8	35.8	30.6	13.3	4.8	1.4	1.0		13.2
	2002	1,351	86.2	38.2	31.2	11.0	3.8	1.2	1.0		13.8
	2003	1,433	88.3	41.9	29.4	11.0	4.4	0.9	0.7		11.7
	2004	1,439	85.5	41.4	28.3	10.0	3.7	1.5	0.7		14.5
	2005	1,341	85.2	42.6	27.8	9.5	4.0	0.7	0.7		14.8
Rest of New Commonwealth	1995	1,332	89.6	44.0	33.2	10.2	1.4	0.6	0.2		10.4
	2002	2,507	94.7	50.4	32.3	9.6	2.0	0.3	0.1		5.3
	2003	2,825	94.4	48.1	34.1	10.1	1.8	0.3	0.1		5.6
	2004	3,228	93.2	47.1	32.8	10.6	1.9	0.6	0.2		6.8
	2005	3,364	93.4	45.6	35.0	10.6	1.5	0.5	0.1		6.6
Rest of the World	1995	32,416	81.4	35.3	27.7	11.2	3.9	1.6	1.5		18.6
	2002	47,666	77.5	35.6	24.6	10.2	3.9	1.5	1.6		22.5
	2003	55,882	76.3	35.5	24.2	9.6	3.9	1.5	1.6		23.7
	2004	61,781	75.8	35.8	24.0	9.2	3.6	1.5	1.6		24.2
	2005	68,344	74.0	35.2	23.3	9.0	3.4	1.6	1.6		26.0

Note: For comparability, the births data for all years for mothers born outside the United Kingdom were reclassified according to the 2005 country classification list and the definition of the European Union (EU25), as constituted in 2005, was used for all years data.
1 Includes births to women whose country of birth was not stated.
2 Including Isle of Man and Channel Islands.
3 See section 2.7.

Table 10.1 Age-specific fertility rates[1]: age and year of birth of woman, 1920-1990

Year of birth of woman/female birth cohort	Age of woman - completed years														
	15	16	17	18	19	20	21	22	23	24	25	26	27	28	29
1920	0	3	10	23	40	59	84	104	117	131	119	159	165	136	123
1925	0	2	9	22	43	67	101	147	157	156	147	141	134	131	118
1926	1	3	9	23	44	72	124	142	151	150	148	148	142	133	121
1927	1	3	10	24	45	86	124	138	144	149	148	152	142	134	128
1928	1	3	11	24	54	88	123	134	146	152	155	150	143	143	135
1929	1	4	10	27	59	90	117	132	146	158	156	151	148	148	137
1930	1	3	11	30	60	87	117	138	155	162	161	165	160	154	141
1931	1	3	12	32	58	85	119	143	156	162	169	170	162	153	146
1932	1	3	14	31	57	87	123	142	154	171	175	173	160	156	147
1933	1	4	13	31	58	90	125	146	167	180	183	178	172	165	151
1934	1	3	12	32	61	92	128	157	175	188	190	189	179	173	155
1935	1	3	13	33	61	94	137	162	180	190	198	193	181	173	158
1936	1	3	13	35	63	103	141	167	179	191	201	196	184	174	151
1937	1	3	14	34	70	107	147	168	184	198	206	199	188	167	142
1938	1	3	15	40	75	112	150	174	194	204	210	202	181	162	135
1939	1	4	18	44	79	116	155	180	200	208	213	193	174	154	131
1940	1	5	20	48	83	123	162	189	204	214	208	191	169	152	127
1941	1	5	22	52	91	127	169	193	209	210	201	183	162	144	123
1942	1	6	24	54	91	132	169	192	202	202	193	178	159	139	121
1943	1	7	27	57	95	132	167	187	197	191	187	172	154	139	112
1944	1	8	31	63	98	135	166	183	188	189	180	168	154	128	105
1945	2	11	35	66	103	136	166	176	181	181	178	168	145	121	101
1946	2	12	36	67	104	136	161	172	175	179	178	159	139	119	100
1947	2	12	36	66	99	128	151	158	165	169	159	145	130	111	97
1948	3	13	38	67	102	127	147	161	167	160	153	142	127	114	100
1949	2	13	38	71	101	124	144	155	152	148	144	134	125	113	106
1950	3	14	41	73	101	124	142	141	142	141	137	131	123	118	114
1951	3	15	43	75	102	125	129	130	133	130	131	126	126	127	115
1952	3	16	45	77	104	115	121	126	125	128	129	132	136	129	114
1953	3	17	46	78	97	107	115	118	121	122	133	141	138	125	114
1954	4	17	47	73	89	99	105	111	114	127	139	142	133	122	113
1955	4	18	47	68	83	92	101	107	119	134	144	135	129	123	114
1956	4	19	42	62	76	86	95	110	126	136	138	133	129	124	117
1957	4	18	41	57	70	81	98	115	128	131	133	134	130	126	116
1958	4	16	35	50	63	81	100	116	119	127	132	133	131	123	118
1959	4	13	30	46	63	83	102	107	116	124	129	132	128	125	119
1960	3	12	28	45	66	85	96	105	114	123	129	128	130	125	117
1961	3	11	27	47	66	78	90	102	110	119	124	129	127	122	120
1962	3	11	28	47	61	74	86	98	108	115	123	126	124	125	117
1963	3	11	27	42	58	71	83	96	105	114	122	122	127	122	117
1964	3	11	24	41	55	68	83	93	104	114	117	123	123	121	116
1965	3	10	24	39	55	69	81	93	104	109	119	120	120	119	115
1966	2	10	24	40	58	70	82	94	101	109	115	119	117	116	113
1967	2	10	25	43	59	71	85	92	100	107	111	114	115	112	111
1968	3	11	27	45	61	74	83	91	99	104	107	111	111	112	111
1969	3	12	28	46	63	72	82	89	94	99	103	106	109	110	109
1970	3	12	28	48	61	72	80	86	91	95	99	104	106	107	106
1971	3	12	30	46	62	71	76	81	85	89	97	101	104	104	103
1972	3	13	29	47	61	68	73	77	81	87	92	97	99	99	99
1973	3	13	31	47	59	66	70	75	81	86	90	94	94	95	99
1974	4	13	30	45	57	65	70	77	80	85	87	90	93	96	106
1975	4	14	30	44	55	63	70	75	78	81	84	87	92	102	109
1976	4	13	30	43	56	67	71	75	76	78	81	87	98	105	111
1977	4	14	28	44	58	66	69	73	73	75	81	91	98	105	
1978	4	13	28	46	58	65	69	70	72	75	84	92	99		
1979	4	13	31	46	58	65	67	69	72	79	88	93			
1980	4	14	31	46	59	63	66	70	75	81	87				
1981	4	14	31	46	57	63	66	70	76	81					
1982	4	14	30	45	56	62	68	74	78						
1983	4	13	29	43	54	63	68	73							
1984	4	12	27	41	54	63	67								
1985	4	11	26	41	53	61									
1986	4	11	26	41	53										
1987	3	11	26	39											
1988	3	11	25												
1989	3	10													
1990	3														

Notes: The age-specific fertility rates refer to 'all live births per 1,000 women' at the age shown.
Live births to women aged under 15 are not included in the calculation of the rate for age 15.
1 See sections 2.11 and 3.9.
2 Includes births at ages 45 and over, achieved up to the end of 2005 by women born in 1960 and earlier years.

30	31	32	33	34	35	36	37	38	39	40	41	42	43	44	45[2]	Year of birth of woman/female birth cohort
109	93	87	76	67	58	52	44	38	32	25	17	13	8	5	4	1920
108	97	90	78	72	65	56	47	40	32	23	15	11	6	3	3	1925
113	100	94	80	76	67	58	48	43	31	22	15	10	5	3	3	1926
119	103	95	85	77	67	57	49	40	29	21	13	9	5	3	2	1927
123	104	101	86	78	67	59	46	38	28	20	12	8	5	2	2	1928
125	109	102	86	79	68	55	43	36	26	18	11	7	4	2	2	1929
136	117	109	91	83	68	54	43	34	24	17	11	7	4	2	2	1930
134	117	107	91	77	63	51	39	30	21	16	10	6	3	2	2	1931
132	116	111	88	72	60	47	36	28	20	13	8	5	3	2	2	1932
137	117	105	85	70	57	44	34	26	18	11	7	4	2	1	2	1933
140	112	97	79	67	53	42	32	22	15	10	6	4	2	1	2	1934
134	104	92	73	62	50	39	28	19	13	9	6	4	2	2	2	1935
127	101	88	70	57	47	34	23	17	12	8	5	4	2	2	2	1936
121	97	83	66	54	41	29	21	16	11	7	5	4	2	1	2	1937
117	94	78	64	47	35	25	19	14	11	8	5	4	2	1	2	1938
111	90	75	56	41	31	24	18	14	10	8	6	4	2	1	2	1939
108	88	67	50	38	29	23	17	14	11	8	7	4	2	1	2	1940
107	80	60	46	36	28	22	18	15	12	8	6	4	2	2	2	1941
96	72	57	43	34	28	23	20	16	12	8	6	4	2	1	2	1942
89	68	54	43	35	30	25	21	15	11	8	6	4	2	1	2	1943
85	66	53	43	37	32	27	21	15	11	8	6	3	2	1	2	1944
83	66	56	47	42	34	26	21	16	12	9	6	4	2	1	2	1945
85	69	60	54	44	34	28	22	17	13	9	6	4	2	1	2	1946
83	73	64	53	41	34	27	22	16	12	9	6	4	2	1	1	1947
92	82	69	56	45	36	29	23	18	13	9	6	4	3	1	1	1948
99	85	69	55	46	37	31	24	19	14	10	7	4	3	1	1	1949
103	85	68	56	47	39	31	26	20	15	11	7	5	3	2	2	1950
100	83	70	58	49	42	34	27	21	16	11	8	5	3	2	2	1951
100	86	72	62	51	44	35	28	22	17	12	8	5	3	2	2	1952
102	88	76	64	54	45	37	30	23	18	13	8	5	3	2	2	1953
103	90	77	67	57	47	39	30	24	18	12	9	6	3	2	2	1954
106	91	81	68	58	49	40	32	24	19	14	9	6	4	2	2	1955
106	94	82	68	61	50	41	33	25	20	14	10	6	4	2	2	1956
109	94	82	72	62	52	42	34	26	20	15	10	6	4	2	3	1957
108	96	86	73	63	52	44	35	27	21	15	10	7	4	2	2	1958
108	99	86	74	63	55	44	35	29	22	16	11	7	4	2	2	1959
113	98	88	75	66	55	47	38	30	23	16	11	7	5	3	3	1960
111	99	87	76	66	56	48	38	29	23	17	11	8	5	3		1961
110	99	90	77	68	59	49	38	30	23	17	12	8	5			1962
109	101	89	79	69	60	49	40	31	24	19	13	9				1963
111	99	91	81	71	61	51	41	32	27	20	14					1964
109	102	92	82	72	61	51	42	35	28	21						1965
108	100	93	82	71	62	53	46	37	29							1966
107	100	91	80	71	63	56	48	38								1967
108	99	90	81	74	69	60	50									1968
105	97	91	83	79	72	61										1969
102	98	93	89	84	75											1970
101	99	99	95	86												1971
101	103	101	95													1972
105	109	103														1973
111	113															1974
111																1975
																1976
																1977
																1978
																1979
																1980
																1981
																1982
																1983
																1984
																1985
																1986
																1987
																1988
																1989
																1990

Table 10.2 **Average number of live-born children[1]: age and year of birth of woman, 1920-1990**

Year of birth of woman/female birth cohort	Age of woman - completed years														
	15	16	17	18	19	20	21	22	23	24	25	26	27	28	29
1920	0.00	0.00	0.01	0.04	0.08	0.13	0.22	0.32	0.44	0.57	0.69	0.85	1.01	1.15	1.27
1925	0.00	0.00	0.01	0.03	0.08	0.14	0.24	0.39	0.55	0.70	0.85	0.99	1.13	1.26	1.37
1926	0.00	0.00	0.01	0.04	0.08	0.15	0.28	0.42	0.57	0.72	0.87	1.01	1.16	1.29	1.41
1927	0.00	0.00	0.01	0.04	0.08	0.17	0.29	0.43	0.57	0.72	0.87	1.02	1.17	1.30	1.43
1928	0.00	0.00	0.02	0.04	0.09	0.18	0.30	0.44	0.58	0.74	0.89	1.04	1.18	1.33	1.46
1929	0.00	0.00	0.01	0.04	0.10	0.19	0.31	0.44	0.59	0.75	0.90	1.05	1.20	1.35	1.49
1930	0.00	0.00	0.01	0.04	0.10	0.19	0.31	0.45	0.60	0.76	0.93	1.09	1.25	1.41	1.55
1931	0.00	0.00	0.02	0.05	0.11	0.19	0.31	0.45	0.61	0.77	0.94	1.11	1.27	1.42	1.57
1932	0.00	0.00	0.02	0.05	0.11	0.19	0.32	0.46	0.61	0.78	0.96	1.13	1.29	1.45	1.59
1933	0.00	0.00	0.02	0.05	0.11	0.20	0.32	0.47	0.63	0.81	1.00	1.18	1.35	1.51	1.66
1934	0.00	0.00	0.02	0.05	0.11	0.20	0.33	0.49	0.66	0.85	1.04	1.23	1.41	1.58	1.73
1935	0.00	0.00	0.02	0.05	0.11	0.20	0.34	0.50	0.68	0.87	1.07	1.26	1.45	1.62	1.78
1936	0.00	0.00	0.02	0.05	0.12	0.22	0.36	0.53	0.71	0.90	1.10	1.29	1.48	1.65	1.80
1937	0.00	0.00	0.02	0.05	0.12	0.23	0.38	0.55	0.73	0.93	1.13	1.33	1.52	1.69	1.83
1938	0.00	0.00	0.02	0.06	0.13	0.25	0.40	0.57	0.76	0.97	1.18	1.38	1.56	1.72	1.86
1939	0.00	0.00	0.02	0.07	0.14	0.26	0.42	0.60	0.80	1.00	1.22	1.41	1.58	1.74	1.87
1940	0.00	0.01	0.03	0.07	0.16	0.28	0.44	0.63	0.83	1.05	1.26	1.45	1.61	1.77	1.89
1941	0.00	0.01	0.03	0.08	0.17	0.30	0.47	0.66	0.87	1.08	1.28	1.46	1.63	1.77	1.89
1942	0.00	0.01	0.03	0.08	0.18	0.31	0.48	0.67	0.87	1.07	1.27	1.44	1.60	1.74	1.86
1943	0.00	0.01	0.04	0.09	0.19	0.32	0.49	0.67	0.87	1.06	1.25	1.42	1.57	1.71	1.83
1944	0.00	0.01	0.04	0.10	0.20	0.34	0.50	0.69	0.87	1.06	1.24	1.41	1.56	1.69	1.80
1945	0.00	0.01	0.05	0.11	0.22	0.35	0.52	0.69	0.88	1.06	1.23	1.40	1.55	1.67	1.77
1946	0.00	0.01	0.05	0.12	0.22	0.36	0.52	0.69	0.86	1.04	1.22	1.38	1.52	1.64	1.74
1947	0.00	0.01	0.05	0.12	0.22	0.34	0.49	0.65	0.82	0.99	1.14	1.29	1.42	1.53	1.63
1948	0.00	0.02	0.05	0.12	0.22	0.35	0.50	0.66	0.83	0.99	1.14	1.28	1.41	1.52	1.62
1949	0.00	0.02	0.05	0.13	0.23	0.35	0.50	0.65	0.80	0.95	1.09	1.23	1.35	1.47	1.57
1950	0.00	0.02	0.06	0.13	0.23	0.36	0.50	0.64	0.78	0.92	1.06	1.19	1.31	1.43	1.54
1951	0.00	0.02	0.06	0.14	0.24	0.36	0.49	0.62	0.76	0.89	1.02	1.14	1.27	1.40	1.51
1952	0.00	0.02	0.07	0.14	0.25	0.36	0.48	0.61	0.73	0.86	0.99	1.12	1.26	1.39	1.50
1953	0.00	0.02	0.07	0.14	0.24	0.35	0.46	0.58	0.70	0.82	0.96	1.10	1.24	1.36	1.48
1954	0.00	0.02	0.07	0.14	0.23	0.33	0.44	0.55	0.66	0.79	0.93	1.07	1.20	1.32	1.44
1955	0.00	0.02	0.07	0.14	0.22	0.31	0.41	0.52	0.64	0.77	0.92	1.05	1.18	1.31	1.42
1956	0.00	0.02	0.07	0.13	0.20	0.29	0.39	0.50	0.62	0.76	0.90	1.03	1.16	1.28	1.40
1957	0.00	0.02	0.06	0.12	0.19	0.27	0.37	0.49	0.61	0.74	0.88	1.01	1.14	1.27	1.38
1958	0.00	0.02	0.06	0.11	0.17	0.25	0.35	0.47	0.58	0.71	0.84	0.98	1.11	1.23	1.35
1959	0.00	0.02	0.05	0.09	0.16	0.24	0.34	0.45	0.56	0.69	0.82	0.95	1.08	1.20	1.32
1960	0.00	0.02	0.04	0.09	0.15	0.24	0.34	0.44	0.55	0.68	0.81	0.93	1.06	1.19	1.31
1961	0.00	0.02	0.04	0.09	0.16	0.23	0.32	0.42	0.53	0.65	0.78	0.91	1.03	1.16	1.28
1962	0.00	0.01	0.04	0.09	0.15	0.23	0.31	0.41	0.52	0.63	0.76	0.88	1.01	1.13	1.25
1963	0.00	0.01	0.04	0.08	0.14	0.21	0.30	0.39	0.50	0.61	0.73	0.86	0.98	1.10	1.22
1964	0.00	0.01	0.04	0.08	0.13	0.20	0.29	0.38	0.48	0.60	0.71	0.84	0.96	1.08	1.20
1965	0.00	0.01	0.04	0.08	0.13	0.20	0.28	0.38	0.48	0.59	0.71	0.83	0.95	1.06	1.18
1966	0.00	0.01	0.04	0.08	0.13	0.20	0.29	0.38	0.48	0.59	0.71	0.82	0.94	1.06	1.17
1967	0.00	0.01	0.04	0.08	0.14	0.21	0.30	0.39	0.49	0.60	0.71	0.82	0.94	1.05	1.16
1968	0.00	0.01	0.04	0.09	0.15	0.22	0.30	0.40	0.49	0.60	0.71	0.82	0.93	1.04	1.15
1969	0.00	0.02	0.04	0.09	0.15	0.22	0.31	0.39	0.49	0.59	0.69	0.80	0.91	1.02	1.12
1970	0.00	0.02	0.04	0.09	0.15	0.22	0.30	0.39	0.48	0.58	0.67	0.78	0.88	0.99	1.10
1971	0.00	0.02	0.05	0.09	0.15	0.22	0.30	0.38	0.47	0.56	0.65	0.75	0.86	0.96	1.06
1972	0.00	0.02	0.05	0.09	0.15	0.22	0.30	0.37	0.45	0.54	0.63	0.73	0.83	0.93	1.03
1973	0.00	0.02	0.05	0.09	0.15	0.22	0.29	0.36	0.44	0.53	0.62	0.71	0.81	0.90	1.00
1974	0.00	0.02	0.05	0.09	0.15	0.22	0.28	0.36	0.44	0.53	0.61	0.70	0.80	0.89	1.00
1975	0.00	0.02	0.05	0.09	0.15	0.21	0.28	0.35	0.43	0.51	0.60	0.69	0.78	0.88	0.99
1976	0.00	0.02	0.05	0.09	0.15	0.21	0.28	0.36	0.44	0.51	0.60	0.68	0.78	0.89	1.00
1977	0.00	0.02	0.05	0.09	0.15	0.22	0.28	0.36	0.43	0.51	0.59	0.68	0.78	0.88	
1978	0.00	0.02	0.05	0.09	0.15	0.21	0.28	0.35	0.42	0.50	0.58	0.68	0.77		
1979	0.00	0.02	0.05	0.09	0.15	0.22	0.28	0.35	0.43	0.51	0.59	0.69			
1980	0.00	0.02	0.05	0.10	0.15	0.22	0.28	0.35	0.43	0.51	0.60				
1981	0.01	0.02	0.05	0.10	0.15	0.22	0.28	0.35	0.43	0.51					
1982	0.01	0.02	0.05	0.09	0.15	0.21	0.28	0.35	0.43						
1983	0.00	0.02	0.05	0.09	0.14	0.21	0.28	0.35							
1984	0.00	0.02	0.04	0.09	0.14	0.20	0.27								
1985	0.00	0.02	0.04	0.08	0.14	0.20									
1986	0.00	0.02	0.04	0.08	0.14										
1987	0.00	0.02	0.04	0.08											
1988	0.00	0.01	0.04												
1989	0.00	0.01													
1990	0.00														

1 See section 3.9.
2 Includes births at ages 45 and over, achieved up to the end of 2005 by women born in 1960 and earlier years.

England and Wales

30	31	32	33	34	35	36	37	38	39	40	41	42	43	44	45[2]	Year of birth of woman/female birth cohort
1.38	1.47	1.56	1.64	1.70	1.76	1.81	1.86	1.90	1.93	1.95	1.97	1.98	1.99	2.00	2.00	1920
1.48	1.58	1.67	1.75	1.82	1.88	1.94	1.99	2.03	2.06	2.08	2.10	2.11	2.11	2.12	2.12	1925
1.52	1.62	1.72	1.80	1.87	1.94	2.00	2.05	2.09	2.12	2.14	2.16	2.17	2.17	2.18	2.18	1926
1.55	1.65	1.74	1.83	1.91	1.97	2.03	2.08	2.12	2.15	2.17	2.18	2.19	2.19	2.20	2.20	1927
1.58	1.69	1.79	1.88	1.95	2.02	2.08	2.13	2.16	2.19	2.21	2.22	2.23	2.24	2.24	2.24	1928
1.61	1.72	1.82	1.91	1.99	2.05	2.11	2.15	2.19	2.21	2.23	2.24	2.25	2.24	2.26	2.26	1929
1.68	1.80	1.91	2.00	2.08	2.15	2.20	2.25	2.28	2.30	2.32	2.33	2.34	2.34	2.34	2.35	1930
1.70	1.82	1.93	2.02	2.10	2.16	2.21	2.25	2.28	2.30	2.32	2.33	2.33	2.34	2.34	2.34	1931
1.73	1.84	1.95	2.04	2.11	2.17	2.22	2.26	2.28	2.30	2.32	2.33	2.33	2.33	2.33	2.34	1932
1.80	1.92	2.02	2.11	2.18	2.23	2.28	2.31	2.34	2.36	2.37	2.37	2.38	2.38	2.38	2.39	1933
1.88	1.99	2.08	2.16	2.23	2.28	2.33	2.36	2.38	2.40	2.41	2.41	2.41	2.42	2.42	2.42	1934
1.91	2.01	2.11	2.18	2.24	2.29	2.33	2.36	2.38	2.39	2.40	2.41	2.41	2.41	2.41	2.42	1935
1.93	2.03	2.12	2.19	2.25	2.29	2.33	2.35	2.37	2.38	2.39	2.39	2.40	2.40	2.40	2.40	1936
1.95	2.05	2.13	2.20	2.25	2.29	2.32	2.34	2.35	2.37	2.37	2.38	2.38	2.38	2.39	2.39	1937
1.98	2.07	2.15	2.21	2.26	2.29	2.32	2.34	2.35	2.36	2.37	2.38	2.38	2.38	2.38	2.39	1938
1.98	2.07	2.15	2.20	2.24	2.27	2.30	2.32	2.33	2.34	2.35	2.35	2.36	2.36	2.36	2.36	1939
2.00	2.09	2.16	2.21	2.24	2.27	2.30	2.31	2.33	2.34	2.35	2.36	2.36	2.36	2.36	2.36	1940
2.00	2.08	2.14	2.19	2.22	2.25	2.27	2.29	2.31	2.32	2.33	2.33	2.33	2.34	2.34	2.34	1941
1.96	2.03	2.09	2.13	2.16	2.19	2.21	2.23	2.25	2.26	2.27	2.28	2.28	2.28	2.28	2.29	1942
1.91	1.98	2.04	2.08	2.12	2.15	2.17	2.19	2.21	2.22	2.23	2.23	2.24	2.24	2.24	2.24	1943
1.88	1.95	2.00	2.04	2.08	2.11	2.14	2.16	2.18	2.19	2.20	2.20	2.20	2.21	2.21	2.21	1944
1.85	1.92	1.97	2.02	2.06	2.10	2.12	2.14	2.16	2.17	2.18	2.19	2.19	2.19	2.19	2.19	1945
1.82	1.89	1.95	2.00	2.05	2.08	2.11	2.13	2.15	2.16	2.17	2.18	2.18	2.18	2.18	2.19	1946
1.71	1.78	1.85	1.90	1.94	1.98	2.00	2.02	2.04	2.05	2.06	2.07	2.07	2.07	2.08	2.08	1947
1.71	1.80	1.86	1.92	1.97	2.00	2.03	2.05	2.07	2.08	2.09	2.10	2.10	2.11	2.11	2.11	1948
1.67	1.76	1.82	1.88	1.93	1.96	1.99	2.02	2.04	2.05	2.06	2.07	2.07	2.07	2.07	2.08	1949
1.65	1.73	1.80	1.86	1.91	1.94	1.98	2.00	2.02	2.04	2.05	2.05	2.06	2.06	2.06	2.07	1950
1.61	1.69	1.77	1.82	1.87	1.91	1.95	1.98	2.00	2.01	2.02	2.03	2.04	2.04	2.04	2.04	1951
1.60	1.69	1.76	1.82	1.87	1.92	1.95	1.98	2.00	2.02	2.03	2.04	2.04	2.05	2.05	2.05	1952
1.58	1.67	1.74	1.81	1.86	1.91	1.94	1.97	1.99	2.00	2.01	2.03	2.03	2.04	2.04	2.05	1953
1.54	1.63	1.71	1.77	1.83	1.88	1.92	1.95	1.97	1.99	2.00	2.01	2.02	2.02	2.02	2.02	1954
1.53	1.62	1.70	1.76	1.82	1.87	1.91	1.94	1.97	1.99	2.00	2.01	2.02	2.02	2.02	2.02	1955
1.51	1.60	1.68	1.75	1.81	1.86	1.90	1.93	1.96	1.98	1.99	2.00	2.01	2.01	2.01	2.02	1956
1.49	1.59	1.67	1.74	1.80	1.85	1.89	1.93	1.95	1.97	1.99	2.00	2.01	2.01	2.01	2.01	1957
1.46	1.55	1.64	1.71	1.77	1.83	1.87	1.90	1.93	1.95	1.97	1.98	1.98	1.99	1.99	1.99	1958
1.43	1.53	1.62	1.69	1.75	1.81	1.85	1.89	1.92	1.94	1.95	1.96	1.97	1.98	1.98	1.98	1959
1.42	1.52	1.60	1.68	1.75	1.80	1.85	1.89	1.92	1.94	1.95	1.96	1.97	1.98	1.98	1.98	1960
1.39	1.49	1.57	1.65	1.71	1.77	1.82	1.86	1.89	1.91	1.93	1.94	1.94	1.95	1.95		1961
1.36	1.46	1.55	1.62	1.69	1.75	1.80	1.84	1.87	1.89	1.91	1.92	1.93	1.93			1962
1.33	1.43	1.52	1.60	1.67	1.73	1.78	1.82	1.85	1.87	1.89	1.91					1963
1.31	1.41	1.50	1.58	1.65	1.71	1.76	1.80	1.83	1.86	1.88						1964
1.29	1.39	1.48	1.56	1.64	1.70	1.75	1.79	1.83	1.85	1.87						1965
1.28	1.38	1.47	1.55	1.62	1.69	1.74	1.79	1.82	1.85							1966
1.27	1.37	1.46	1.54	1.61	1.67	1.73	1.77	1.81								1967
1.26	1.36	1.45	1.53	1.60	1.67	1.73	1.78									1968
1.23	1.33	1.42	1.50	1.58	1.65	1.71										1969
1.20	1.30	1.39	1.48	1.56	1.64											1970
1.17	1.26	1.36	1.46	1.54												1971
1.13	1.23	1.33	1.43													1972
1.11	1.22	1.32														1973
1.11	1.22															1974
1.10																1975
																1976
																1977
																1978
																1979
																1980
																1981
																1982
																1983
																1984
																1985
																1986
																1987
																1988
																1989
																1990

Table 10.3 Estimated average number of first live-born children[1]:
age and year of birth of woman, 1920-1990

Year of birth of woman/female birth cohort	Age of woman - completed years														
	15	16	17	18	19	20	21	22	23	24	25	26	27	28	29
1920	0.00	0.00	0.01	0.03	0.07	0.11	0.17	0.25	0.32	0.40	0.45	0.53	0.60	0.64	0.67
1925	0.00	0.00	0.01	0.03	0.07	0.12	0.20	0.31	0.40	0.48	0.54	0.60	0.64	0.68	0.71
1926	0.00	0.00	0.01	0.03	0.07	0.13	0.23	0.32	0.41	0.48	0.54	0.60	0.65	0.68	0.72
1927	0.00	0.00	0.01	0.03	0.07	0.15	0.24	0.33	0.41	0.48	0.54	0.60	0.65	0.69	0.72
1928	0.00	0.00	0.01	0.04	0.08	0.15	0.24	0.33	0.41	0.48	0.55	0.61	0.65	0.69	0.73
1929	0.00	0.00	0.01	0.04	0.09	0.16	0.24	0.32	0.40	0.48	0.55	0.61	0.65	0.70	0.73
1930	0.00	0.00	0.01	0.04	0.09	0.16	0.24	0.33	0.41	0.49	0.56	0.62	0.67	0.72	0.75
1931	0.00	0.00	0.02	0.04	0.09	0.16	0.24	0.33	0.41	0.49	0.56	0.62	0.68	0.72	0.75
1932	0.00	0.00	0.02	0.04	0.09	0.16	0.24	0.33	0.41	0.50	0.57	0.63	0.68	0.72	0.75
1933	0.00	0.00	0.02	0.04	0.09	0.16	0.25	0.34	0.43	0.51	0.59	0.65	0.70	0.74	0.77
1934	0.00	0.00	0.02	0.04	0.09	0.17	0.25	0.35	0.45	0.53	0.61	0.67	0.72	0.76	0.79
1935	0.00	0.00	0.02	0.04	0.09	0.17	0.26	0.36	0.45	0.54	0.61	0.67	0.72	0.77	0.80
1936	0.00	0.00	0.02	0.05	0.10	0.18	0.27	0.37	0.46	0.54	0.62	0.68	0.73	0.77	0.80
1937	0.00	0.00	0.02	0.05	0.10	0.18	0.28	0.38	0.47	0.55	0.62	0.69	0.73	0.77	0.80
1938	0.00	0.00	0.02	0.05	0.11	0.20	0.29	0.39	0.48	0.57	0.64	0.70	0.75	0.78	0.81
1939	0.00	0.00	0.02	0.06	0.12	0.21	0.30	0.40	0.50	0.58	0.65	0.71	0.75	0.79	0.81
1940	0.00	0.01	0.02	0.07	0.13	0.22	0.32	0.42	0.51	0.59	0.66	0.72	0.76	0.79	0.82
1941	0.00	0.01	0.03	0.07	0.14	0.23	0.33	0.43	0.52	0.60	0.67	0.73	0.77	0.80	0.82
1942	0.00	0.01	0.03	0.07	0.14	0.23	0.33	0.43	0.52	0.60	0.67	0.72	0.76	0.79	0.82
1943	0.00	0.01	0.03	0.08	0.15	0.24	0.33	0.43	0.52	0.59	0.66	0.71	0.75	0.79	0.81
1944	0.00	0.01	0.04	0.09	0.16	0.25	0.34	0.44	0.52	0.60	0.66	0.71	0.76	0.79	0.82
1945	0.00	0.01	0.04	0.10	0.17	0.26	0.35	0.44	0.53	0.60	0.66	0.72	0.76	0.80	0.82
1946	0.00	0.01	0.05	0.10	0.17	0.26	0.35	0.44	0.52	0.59	0.66	0.71	0.76	0.79	0.82
1947	0.00	0.01	0.05	0.10	0.17	0.25	0.34	0.42	0.49	0.56	0.63	0.68	0.72	0.76	0.79
1948	0.00	0.02	0.05	0.10	0.18	0.26	0.34	0.42	0.50	0.56	0.62	0.68	0.72	0.76	0.79
1949	0.00	0.02	0.05	0.11	0.18	0.26	0.34	0.41	0.48	0.55	0.60	0.66	0.70	0.74	0.77
1950	0.00	0.02	0.05	0.11	0.19	0.26	0.34	0.41	0.47	0.53	0.59	0.64	0.68	0.72	0.76
1951	0.00	0.02	0.06	0.12	0.19	0.27	0.34	0.40	0.46	0.52	0.57	0.62	0.67	0.71	0.74
1952	0.00	0.02	0.06	0.12	0.20	0.27	0.33	0.39	0.45	0.50	0.55	0.61	0.66	0.70	0.73
1953	0.00	0.02	0.06	0.12	0.19	0.26	0.32	0.38	0.43	0.48	0.54	0.60	0.65	0.69	0.72
1954	0.00	0.02	0.06	0.12	0.18	0.24	0.30	0.35	0.41	0.46	0.52	0.58	0.63	0.67	0.70
1955	0.00	0.02	0.06	0.12	0.18	0.23	0.29	0.34	0.40	0.46	0.52	0.57	0.62	0.66	0.69
1956	0.00	0.02	0.06	0.11	0.16	0.22	0.27	0.32	0.38	0.45	0.50	0.55	0.60	0.64	0.68
1957	0.00	0.02	0.06	0.11	0.15	0.20	0.26	0.32	0.38	0.44	0.49	0.54	0.59	0.63	0.67
1958	0.00	0.02	0.05	0.09	0.14	0.19	0.24	0.31	0.36	0.42	0.47	0.53	0.58	0.62	0.66
1959	0.00	0.02	0.04	0.08	0.13	0.18	0.24	0.29	0.35	0.40	0.46	0.51	0.56	0.60	0.65
1960	0.00	0.02	0.04	0.08	0.13	0.18	0.23	0.29	0.34	0.40	0.45	0.50	0.55	0.60	0.64
1961	0.00	0.01	0.04	0.08	0.13	0.18	0.23	0.28	0.33	0.38	0.44	0.49	0.54	0.59	0.63
1962	0.00	0.01	0.04	0.08	0.12	0.17	0.22	0.27	0.32	0.37	0.43	0.48	0.53	0.58	0.62
1963	0.00	0.01	0.04	0.07	0.12	0.16	0.21	0.26	0.31	0.37	0.42	0.47	0.52	0.57	0.61
1964	0.00	0.01	0.04	0.07	0.11	0.16	0.20	0.25	0.31	0.36	0.41	0.46	0.51	0.56	0.60
1965	0.00	0.01	0.04	0.07	0.11	0.15	0.20	0.25	0.30	0.35	0.41	0.46	0.51	0.55	0.60
1966	0.00	0.01	0.03	0.07	0.11	0.16	0.20	0.25	0.30	0.35	0.41	0.46	0.51	0.55	0.59
1967	0.00	0.01	0.04	0.07	0.12	0.16	0.21	0.26	0.31	0.36	0.41	0.45	0.50	0.55	0.59
1968	0.00	0.01	0.04	0.08	0.12	0.17	0.22	0.27	0.31	0.36	0.41	0.46	0.50	0.55	0.59
1969	0.00	0.01	0.04	0.08	0.13	0.17	0.22	0.27	0.31	0.35	0.40	0.45	0.49	0.54	0.58
1970	0.00	0.02	0.04	0.08	0.13	0.18	0.22	0.27	0.31	0.35	0.39	0.44	0.48	0.53	0.57
1971	0.00	0.02	0.04	0.08	0.13	0.18	0.22	0.26	0.30	0.34	0.38	0.43	0.47	0.52	0.56
1972	0.00	0.02	0.04	0.08	0.13	0.17	0.22	0.25	0.29	0.33	0.37	0.42	0.46	0.50	0.54
1973	0.00	0.02	0.05	0.09	0.13	0.17	0.21	0.25	0.28	0.32	0.36	0.41	0.45	0.49	0.53
1974	0.00	0.02	0.05	0.08	0.13	0.17	0.21	0.24	0.28	0.32	0.36	0.40	0.44	0.48	0.53
1975	0.00	0.02	0.04	0.08	0.12	0.16	0.20	0.24	0.28	0.31	0.35	0.39	0.43	0.47	0.52
1976	0.00	0.02	0.04	0.08	0.12	0.16	0.20	0.24	0.28	0.31	0.35	0.39	0.43	0.48	0.52
1977	0.00	0.02	0.04	0.08	0.12	0.16	0.20	0.24	0.27	0.31	0.35	0.39	0.43	0.48	
1978	0.00	0.02	0.04	0.08	0.12	0.17	0.20	0.24	0.27	0.31	0.35	0.39	0.43		
1979	0.00	0.02	0.05	0.08	0.13	0.17	0.20	0.24	0.27	0.31	0.35	0.39			
1980	0.00	0.02	0.05	0.09	0.13	0.17	0.20	0.24	0.28	0.32	0.36				
1981	0.01	0.02	0.05	0.09	0.13	0.17	0.20	0.24	0.28	0.32					
1982	0.01	0.02	0.05	0.08	0.13	0.17	0.20	0.24	0.28						
1983	0.00	0.02	0.04	0.08	0.12	0.16	0.20	0.23							
1984	0.00	0.02	0.04	0.08	0.11	0.15	0.19								
1985	0.00	0.02	0.04	0.07	0.11	0.15									
1986	0.00	0.01	0.04	0.07	0.11										
1987	0.00	0.01	0.04	0.07											
1988	0.00	0.01	0.04												
1989	0.00	0.01													
1990	0.00														

1 See sections 2.9 and 3.9.
2 Includes births at ages 45 and over, achieved up to the end of 2005 by women born in 1960 and earlier years.

30	31	32	33	34	35	36	37	38	39	40	41	42	43	44	45²	Year of birth of woman/female birth cohort
0.70	0.72	0.74	0.75	0.76	0.77	0.77	0.78	0.78	0.79	0.79	0.79	0.79	0.79	0.79	0.79	1920
0.74	0.76	0.78	0.79	0.80	0.81	0.82	0.82	0.82	0.83	0.83	0.83	0.83	0.83	0.83	0.83	1925
0.74	0.77	0.78	0.80	0.81	0.82	0.82	0.83	0.83	0.83	0.83	0.84	0.84	0.84	0.84	0.84	1926
0.75	0.77	0.79	0.80	0.81	0.82	0.83	0.83	0.83	0.84	0.84	0.84	0.84	0.84	0.84	0.84	1927
0.76	0.78	0.80	0.81	0.82	0.83	0.84	0.84	0.84	0.85	0.85	0.85	0.85	0.85	0.85	0.85	1928
0.76	0.78	0.80	0.81	0.82	0.83	0.84	0.84	0.84	0.85	0.85	0.85	0.85	0.85	0.85	0.85	1929
0.78	0.80	0.82	0.83	0.84	0.85	0.85	0.86	0.86	0.86	0.86	0.86	0.87	0.87	0.87	0.87	1930
0.78	0.80	0.82	0.83	0.84	0.84	0.85	0.85	0.86	0.86	0.86	0.86	0.86	0.86	0.86	0.86	1931
0.78	0.80	0.82	0.83	0.84	0.84	0.85	0.85	0.86	0.86	0.86	0.86	0.86	0.86	0.86	0.86	1932
0.80	0.82	0.83	0.84	0.85	0.86	0.86	0.86	0.87	0.87	0.87	0.87	0.87	0.87	0.87	0.87	1933
0.82	0.84	0.85	0.86	0.87	0.87	0.88	0.88	0.88	0.88	0.89	0.89	0.89	0.89	0.89	0.89	1934
0.82	0.84	0.85	0.86	0.87	0.87	0.87	0.88	0.88	0.88	0.88	0.88	0.88	0.88	0.88	0.88	1935
0.82	0.84	0.85	0.86	0.86	0.87	0.87	0.88	0.88	0.88	0.88	0.88	0.88	0.88	0.88	0.88	1936
0.82	0.84	0.85	0.85	0.86	0.87	0.87	0.87	0.88	0.88	0.88	0.88	0.88	0.88	0.88	0.88	1937
0.83	0.84	0.85	0.86	0.87	0.87	0.88	0.88	0.88	0.88	0.88	0.89	0.89	0.89	0.89	0.89	1938
0.83	0.84	0.85	0.86	0.87	0.87	0.88	0.88	0.88	0.88	0.88	0.88	0.88	0.88	0.88	0.88	1939
0.83	0.85	0.86	0.87	0.87	0.88	0.88	0.88	0.88	0.88	0.89	0.89	0.89	0.89	0.89	0.89	1940
0.84	0.85	0.86	0.87	0.88	0.88	0.89	0.89	0.89	0.89	0.89	0.89	0.89	0.89	0.89	0.89	1941
0.84	0.85	0.86	0.87	0.87	0.88	0.88	0.88	0.89	0.89	0.89	0.89	0.89	0.89	0.89	0.89	1942
0.83	0.85	0.86	0.86	0.87	0.87	0.88	0.88	0.88	0.88	0.88	0.89	0.89	0.89	0.89	0.89	1943
0.84	0.85	0.87	0.87	0.88	0.88	0.89	0.89	0.89	0.90	0.90	0.90	0.90	0.90	0.90	0.90	1944
0.84	0.86	0.87	0.88	0.89	0.89	0.90	0.90	0.90	0.90	0.90	0.90	0.90	0.90	0.90	0.90	1945
0.84	0.86	0.87	0.88	0.89	0.90	0.90	0.90	0.91	0.91	0.91	0.91	0.91	0.91	0.91	0.91	1946
0.81	0.83	0.84	0.85	0.86	0.86	0.87	0.87	0.87	0.87	0.88	0.88	0.88	0.88	0.88	0.88	1947
0.81	0.83	0.85	0.86	0.86	0.87	0.88	0.88	0.88	0.88	0.89	0.89	0.89	0.89	0.89	0.89	1948
0.80	0.81	0.83	0.84	0.85	0.85	0.86	0.86	0.87	0.87	0.87	0.87	0.87	0.87	0.87	0.87	1949
0.78	0.80	0.82	0.83	0.84	0.84	0.85	0.85	0.86	0.86	0.86	0.86	0.86	0.87	0.87	0.87	1950
0.77	0.79	0.80	0.82	0.82	0.83	0.84	0.84	0.85	0.85	0.85	0.85	0.85	0.86	0.86	0.86	1951
0.76	0.78	0.80	0.81	0.82	0.83	0.84	0.84	0.85	0.85	0.85	0.85	0.85	0.85	0.85	0.85	1952
0.75	0.77	0.79	0.80	0.82	0.83	0.83	0.84	0.84	0.85	0.85	0.85	0.85	0.85	0.85	0.85	1953
0.73	0.76	0.77	0.79	0.80	0.81	0.82	0.83	0.83	0.83	0.84	0.84	0.84	0.84	0.84	0.84	1954
0.73	0.75	0.77	0.79	0.80	0.81	0.82	0.83	0.83	0.83	0.84	0.84	0.84	0.84	0.84	0.84	1955
0.71	0.74	0.76	0.78	0.79	0.80	0.81	0.82	0.82	0.83	0.83	0.83	0.83	0.83	0.83	0.84	1956
0.71	0.73	0.75	0.77	0.79	0.80	0.81	0.81	0.82	0.82	0.83	0.83	0.83	0.83	0.83	0.83	1957
0.69	0.72	0.74	0.76	0.77	0.79	0.80	0.80	0.81	0.81	0.82	0.82	0.82	0.82	0.82	0.82	1958
0.68	0.71	0.73	0.75	0.77	0.78	0.79	0.80	0.80	0.81	0.81	0.81	0.82	0.82	0.82	0.82	1959
0.67	0.70	0.73	0.75	0.76	0.77	0.78	0.79	0.80	0.80	0.81	0.81	0.81	0.81	0.81	0.82	1960
0.66	0.69	0.72	0.74	0.75	0.77	0.78	0.79	0.79	0.80	0.80	0.81	0.81	0.81	0.81		1961
0.66	0.69	0.71	0.73	0.75	0.76	0.78	0.78	0.79	0.80	0.80	0.81	0.81	0.81			1962
0.65	0.68	0.71	0.73	0.75	0.76	0.77	0.78	0.79	0.80	0.80	0.81	0.81				1963
0.64	0.67	0.70	0.72	0.74	0.75	0.77	0.78	0.79	0.79	0.80	0.80					1964
0.64	0.67	0.70	0.72	0.74	0.76	0.77	0.78	0.79	0.80	0.80						1965
0.63	0.67	0.70	0.72	0.74	0.75	0.77	0.78	0.79	0.80							1966
0.63	0.66	0.69	0.72	0.73	0.75	0.77	0.78	0.79								1967
0.63	0.67	0.70	0.72	0.74	0.76	0.78	0.79									1968
0.62	0.65	0.68	0.71	0.73	0.75	0.77										1969
0.61	0.64	0.67	0.70	0.73	0.75											1970
0.59	0.63	0.67	0.70	0.72												1971
0.58	0.62	0.65	0.68													1972
0.57	0.61	0.65														1973
0.57	0.62															1974
0.56																1975
																1976
																1977
																1978
																1979
																1980
																1981
																1982
																1983
																1984
																1985
																1986
																1987
																1988
																1989
																1990

Table 10.4   Components of average family size[1]: occurrence   England and Wales
              within/outside marriage, birth order, mother's year
              of birth, and age of mother at birth, 1920-1986

Mother's year of birth	All live births	Outside marriage	Within marriage					
			Birth order					
			All	First	Second	Third	Fourth	Fifth and later
All ages of mothers at birth								
1920	**2.00**	0.13	1.87	0.76	0.55	0.28	0.14	0.15
1925	**2.12**	0.12	2.00	0.79	0.58	0.31	0.15	0.17
1930	**2.35**	0.12	2.23	0.84	0.66	0.36	0.18	0.19
1935	**2.42**	0.13	2.29	0.85	0.71	0.38	0.18	0.16
1940	**2.36**	0.16	2.21	0.85	0.73	0.37	0.16	0.11
1941	**2.34**	0.16	2.18	0.85	0.73	0.36	0.15	0.10
1942	**2.29**	0.16	2.12	0.84	0.72	0.34	0.13	0.08
1943	**2.24**	0.16	2.08	0.84	0.72	0.33	0.12	0.07
1944	**2.21**	0.16	2.04	0.84	0.72	0.31	0.11	0.06
1945	**2.19**	0.17	2.02	0.85	0.72	0.30	0.10	0.05
1946	**2.19**	0.18	2.01	0.85	0.72	0.29	0.10	0.05
1947	**2.08**	0.17	1.91	0.82	0.69	0.27	0.09	0.04
1948	**2.11**	0.18	1.93	0.82	0.71	0.27	0.09	0.04
1949	**2.08**	0.19	1.89	0.81	0.69	0.27	0.08	0.04
1950	**2.07**	0.20	1.87	0.80	0.69	0.26	0.08	0.04
1951	**2.04**	0.20	1.84	0.78	0.67	0.26	0.08	0.04
1952	**2.05**	0.21	1.84	0.78	0.67	0.26	0.08	0.05
1953	**2.05**	0.22	1.82	0.78	0.66	0.26	0.08	0.05
1954	**2.02**	0.23	1.79	0.76	0.65	0.25	0.08	0.05
1955	**2.02**	0.25	1.77	0.75	0.65	0.25	0.08	0.05
1956	**2.02**	0.27	1.75	0.74	0.64	0.25	0.08	0.05
1957	**2.01**	0.29	1.73	0.72	0.63	0.25	0.08	0.05
1958	**1.99**	0.31	1.68	0.71	0.61	0.24	0.08	0.04
1959	**1.98**	0.33	1.65	0.69	0.60	0.24	0.08	0.04
1960	**1.98**	0.36	1.62	0.68	0.59	0.24	0.08	0.04
Under 20								
1920	**0.08**	0.01	0.06	0.06	0.01	0.00		
1925	**0.08**	0.02	0.06	0.05	0.01	0.00		
1930	**0.10**	0.02	0.09	0.08	0.01	0.00		
1935	**0.11**	0.02	0.09	0.08	0.01	0.00		
1940	**0.16**	0.03	0.13	0.11	0.02	0.00		
1941	**0.17**	0.03	0.14	0.12	0.02	0.00		
1942	**0.18**	0.03	0.14	0.12	0.02	0.00		
1943	**0.19**	0.03	0.15	0.12	0.02	0.00		
1944	**0.20**	0.04	0.16	0.13	0.03	0.00		
1945	**0.22**	0.04	0.17	0.14	0.03	0.00		
1946	**0.22**	0.05	0.17	0.14	0.03	0.00		
1947	**0.22**	0.05	0.17	0.13	0.03	0.00		
1948	**0.22**	0.05	0.17	0.13	0.03	0.00		
1949	**0.23**	0.06	0.17	0.13	0.03	0.00		
1950	**0.23**	0.06	0.17	0.14	0.03	0.00		
1951	**0.24**	0.06	0.18	0.14	0.03	0.00		
1952	**0.25**	0.06	0.18	0.15	0.03	0.00		
1953	**0.24**	0.06	0.18	0.14	0.03	0.00		
1954	**0.23**	0.06	0.17	0.14	0.03	0.00		
1955	**0.22**	0.06	0.16	0.13	0.03	0.00		
1956	**0.20**	0.06	0.14	0.11	0.03	0.00		
1957	**0.19**	0.06	0.13	0.10	0.02	0.00		
1958	**0.17**	0.06	0.11	0.09	0.02	0.00		
1959	**0.16**	0.06	0.10	0.08	0.02	0.00		
1960	**0.15**	0.06	0.09	0.08	0.02	0.00		
1961	**0.16**	0.06	0.09	0.08	0.02	0.00		
1962	**0.15**	0.07	0.09	0.07	0.02	0.00		
1963	**0.14**	0.07	0.07	0.06	0.01	0.00		
1964	**0.13**	0.07	0.06	0.05	0.01	0.00		
1965	**0.13**	0.08	0.06	0.04	0.01	0.00		
1966	**0.13**	0.08	0.05	0.04	0.01	0.00		
1967	**0.14**	0.09	0.05	0.04	0.01	0.00		
1968	**0.15**	0.10	0.04	0.03	0.01	0.00		
1969	**0.15**	0.11	0.04	0.03	0.01	0.00		
1970	**0.15**	0.12	0.04	0.03	0.01	0.00		

Note: Average family sizes are obtained by summing rates for each single year of age.
1 See sections 2.9, 2.11 and 3.9.

Table 10.4 - *continued*

Mother's year of birth	All live births	Outside marriage	Within marriage					
			Birth order					
			All	First	Second	Third	Fourth	Fifth and later
	Under 20 - *continued*							
1971	**0.15**	0.12	0.03	0.02	0.01	0.00		
1972	**0.15**	0.13	0.03	0.02	0.01	0.00		
1973	**0.15**	0.13	0.03	0.02	0.01	0.00		
1974	**0.15**	0.13	0.02	0.02	0.01	0.00		
1975	**0.15**	0.13	0.02	0.02	0.00	0.00		
1976	**0.15**	0.13	0.02	0.02	0.00	0.00		
1977	**0.15**	0.13	0.02	0.02	0.00	0.00		
1978	**0.15**	0.13	0.02	0.01	0.00	0.00		
1979	**0.15**	0.14	0.02	0.01	0.00	0.00		
1980	**0.15**	0.14	0.02	0.01	0.00	0.00		
1981	**0.15**	0.14	0.02	0.01	0.00	0.00		
1982	**0.15**	0.13	0.02	0.01	0.00	0.00		
1983	**0.14**	0.13	0.01	0.01	0.00	0.00		
1984	**0.14**	0.13	0.01	0.01	0.00	0.00		
1985	**0.14**	0.12	0.01	0.01	0.00	0.00		
1986	**0.14**	0.12	0.01	0.01	0.00	0.00		
	20-24							
1920	**0.49**	0.04	0.45	0.31	0.11	0.03	0.01	0.00
1925	**0.63**	0.05	0.58	0.38	0.15	0.04	0.01	0.00
1930	**0.66**	0.03	0.63	0.38	0.18	0.05	0.01	0.00
1935	**0.76**	0.04	0.72	0.43	0.21	0.06	0.02	0.00
1940	**0.89**	0.06	0.83	0.44	0.27	0.09	0.03	0.01
1941	**0.91**	0.06	0.84	0.44	0.28	0.09	0.03	0.01
1942	**0.90**	0.07	0.83	0.43	0.28	0.09	0.02	0.01
1943	**0.87**	0.06	0.81	0.42	0.27	0.09	0.02	0.01
1944	**0.86**	0.07	0.79	0.41	0.27	0.08	0.02	0.01
1945	**0.84**	0.07	0.77	0.40	0.27	0.08	0.02	0.01
1946	**0.82**	0.07	0.76	0.39	0.26	0.08	0.02	0.00
1947	**0.77**	0.06	0.71	0.37	0.24	0.07	0.02	0.00
1948	**0.76**	0.06	0.70	0.37	0.25	0.07	0.02	0.00
1949	**0.72**	0.06	0.66	0.35	0.23	0.06	0.01	0.00
1950	**0.69**	0.06	0.63	0.33	0.23	0.06	0.01	0.00
1951	**0.65**	0.05	0.59	0.31	0.21	0.05	0.01	0.00
1952	**0.62**	0.05	0.56	0.29	0.21	0.05	0.01	0.00
1953	**0.58**	0.05	0.53	0.28	0.20	0.05	0.01	0.00
1954	**0.56**	0.05	0.51	0.26	0.19	0.04	0.01	0.00
1955	**0.55**	0.05	0.50	0.26	0.18	0.04	0.01	0.00
1956	**0.55**	0.06	0.50	0.26	0.18	0.04	0.01	0.00
1957	**0.55**	0.06	0.49	0.26	0.17	0.04	0.01	0.00
1958	**0.54**	0.07	0.47	0.25	0.17	0.04	0.01	0.00
1959	**0.53**	0.08	0.45	0.25	0.16	0.04	0.01	0.00
1960	**0.52**	0.09	0.44	0.23	0.15	0.04	0.01	0.00
1961	**0.50**	0.09	0.41	0.22	0.14	0.04	0.01	0.00
1962	**0.48**	0.10	0.38	0.20	0.13	0.03	0.01	0.00
1963	**0.47**	0.11	0.35	0.19	0.12	0.03	0.01	0.00
1964	**0.46**	0.13	0.33	0.18	0.11	0.03	0.01	0.00
1965	**0.46**	0.14	0.31	0.17	0.11	0.03	0.01	0.00
1966	**0.46**	0.16	0.30	0.16	0.10	0.03	0.01	0.00
1967	**0.45**	0.18	0.28	0.15	0.09	0.03	0.01	0.00
1968	**0.45**	0.19	0.26	0.14	0.09	0.03	0.01	0.00
1969	**0.44**	0.20	0.24	0.13	0.08	0.02	0.00	0.00
1970	**0.42**	0.20	0.22	0.12	0.08	0.02	0.00	0.00
1971	**0.40**	0.21	0.20	0.10	0.07	0.02	0.00	0.00
1972	**0.39**	0.21	0.18	0.10	0.06	0.02	0.00	0.00
1973	**0.38**	0.21	0.17	0.09	0.06	0.02	0.00	0.00
1974	**0.38**	0.22	0.16	0.08	0.06	0.02	0.00	0.00
1975	**0.37**	0.22	0.15	0.08	0.05	0.02	0.00	0.00
1976	**0.37**	0.22	0.14	0.08	0.05	0.01	0.00	0.00
1977	**0.36**	0.22	0.14	0.07	0.05	0.01	0.00	0.00
1978	**0.35**	0.22	0.13	0.07	0.04	0.01	0.00	0.00
1979	**0.35**	0.22	0.13	0.07	0.04	0.01	0.00	0.00
1980	**0.36**	0.23	0.13	0.07	0.04	0.01	0.00	0.00
1981	**0.36**	0.23	0.12	0.07	0.04	0.01	0.00	0.00

Table 10.4 - *continued*

Mother's year of birth	All live births	Outside marriage	Within marriage					
			Birth order					
			All	First	Second	Third	Fourth	Fifth and later
25-29								
1920	0.70	0.04	0.66	0.28	0.24	0.10	0.04	0.02
1925	0.67	0.02	0.65	0.23	0.24	0.11	0.04	0.03
1930	0.78	0.02	0.76	0.26	0.27	0.13	0.06	0.04
1935	0.90	0.04	0.87	0.26	0.32	0.17	0.07	0.05
1940	0.85	0.04	0.80	0.22	0.31	0.16	0.07	0.04
1941	0.81	0.04	0.77	0.22	0.30	0.16	0.06	0.04
1942	0.79	0.04	0.75	0.22	0.29	0.15	0.06	0.03
1943	0.76	0.04	0.73	0.22	0.29	0.14	0.05	0.03
1944	0.73	0.03	0.70	0.22	0.29	0.13	0.04	0.02
1945	0.71	0.03	0.68	0.22	0.28	0.12	0.04	0.02
1946	0.69	0.03	0.66	0.23	0.28	0.11	0.03	0.01
1947	0.64	0.03	0.62	0.22	0.26	0.09	0.03	0.01
1948	0.64	0.03	0.61	0.22	0.26	0.09	0.03	0.01
1949	0.62	0.03	0.59	0.22	0.25	0.08	0.02	0.01
1950	0.62	0.03	0.59	0.22	0.25	0.09	0.02	0.01
1951	0.63	0.03	0.59	0.23	0.25	0.09	0.02	0.01
1952	0.64	0.04	0.60	0.23	0.25	0.09	0.03	0.01
1953	0.65	0.04	0.61	0.24	0.25	0.09	0.03	0.01
1954	0.65	0.04	0.61	0.23	0.24	0.09	0.03	0.01
1955	0.65	0.05	0.60	0.23	0.24	0.09	0.03	0.01
1956	0.64	0.05	0.59	0.22	0.24	0.09	0.03	0.01
1957	0.64	0.06	0.58	0.22	0.23	0.09	0.03	0.01
1958	0.64	0.07	0.57	0.22	0.22	0.09	0.03	0.01
1959	0.63	0.07	0.56	0.22	0.22	0.08	0.02	0.01
1960	0.63	0.08	0.55	0.22	0.21	0.08	0.03	0.01
1961	0.62	0.09	0.53	0.21	0.20	0.08	0.02	0.01
1962	0.61	0.10	0.51	0.21	0.19	0.07	0.02	0.01
1963	0.61	0.11	0.50	0.21	0.19	0.07	0.02	0.01
1964	0.60	0.12	0.48	0.20	0.18	0.06	0.02	0.01
1965	0.59	0.13	0.46	0.20	0.17	0.06	0.02	0.01
1966	0.58	0.14	0.44	0.19	0.16	0.06	0.02	0.01
1967	0.56	0.15	0.42	0.18	0.15	0.06	0.02	0.01
1968	0.55	0.15	0.40	0.18	0.15	0.05	0.02	0.01
1969	0.54	0.16	0.38	0.17	0.14	0.05	0.02	0.01
1970	0.52	0.16	0.36	0.17	0.13	0.05	0.02	0.01
1971	0.51	0.17	0.34	0.16	0.12	0.04	0.01	0.01
1972	0.49	0.16	0.32	0.15	0.11	0.04	0.01	0.01
1973	0.47	0.16	0.31	0.15	0.11	0.04	0.01	0.01
1974	0.47	0.17	0.30	0.15	0.10	0.04	0.01	0.00
1975	0.47	0.17	0.30	0.14	0.10	0.04	0.01	0.01
1976	0.48	0.18	0.30	0.15	0.10	0.04	0.01	0.01
30-34								
1920	0.43	0.02	0.41	0.09	0.14	0.09	0.05	0.05
1925	0.45	0.02	0.43	0.09	0.14	0.10	0.05	0.06
1930	0.53	0.02	0.51	0.09	0.15	0.12	0.07	0.08
1935	0.47	0.02	0.44	0.07	0.13	0.11	0.06	0.06
1940	0.35	0.02	0.33	0.06	0.11	0.09	0.05	0.04
1941	0.33	0.02	0.31	0.06	0.10	0.08	0.04	0.03
1942	0.30	0.02	0.29	0.06	0.10	0.07	0.03	0.02
1943	0.29	0.02	0.27	0.06	0.10	0.07	0.03	0.02
1944	0.28	0.02	0.27	0.06	0.11	0.06	0.03	0.02
1945	0.29	0.02	0.28	0.06	0.11	0.07	0.03	0.01
1946	0.31	0.02	0.29	0.07	0.12	0.07	0.03	0.01
1947	0.31	0.02	0.30	0.07	0.12	0.07	0.02	0.01
1948	0.34	0.02	0.32	0.08	0.13	0.08	0.03	0.01
1949	0.35	0.02	0.33	0.08	0.13	0.08	0.03	0.01
1950	0.36	0.02	0.34	0.08	0.14	0.08	0.03	0.01

Note: Average family sizes are obtained by summing rates for each single year of age.

Table 10.4 - *continued*

Mother's year of birth	All live births	Outside marriage	Within marriage					
			Birth order					
			All	First	Second	Third	Fourth	Fifth and later
30-34 - *continued*								
1951	**0.36**	0.03	0.33	0.08	0.13	0.08	0.03	0.02
1952	**0.37**	0.03	0.34	0.08	0.14	0.08	0.03	0.02
1953	**0.38**	0.03	0.35	0.09	0.14	0.08	0.03	0.02
1954	**0.39**	0.04	0.36	0.09	0.14	0.08	0.03	0.02
1955	**0.40**	0.04	0.36	0.10	0.14	0.08	0.03	0.02
1956	**0.41**	0.05	0.36	0.10	0.14	0.08	0.03	0.02
1957	**0.42**	0.06	0.36	0.10	0.14	0.08	0.03	0.02
1958	**0.43**	0.06	0.37	0.10	0.14	0.08	0.03	0.02
1959	**0.43**	0.07	0.37	0.10	0.14	0.08	0.03	0.02
1960	**0.44**	0.07	0.37	0.11	0.15	0.07	0.03	0.01
1961	**0.44**	0.08	0.36	0.11	0.14	0.07	0.02	0.01
1962	**0.44**	0.08	0.36	0.11	0.14	0.07	0.02	0.01
1963	**0.45**	0.09	0.36	0.11	0.15	0.07	0.02	0.01
1964	**0.45**	0.10	0.36	0.11	0.15	0.07	0.02	0.01
1965	**0.46**	0.10	0.36	0.12	0.15	0.06	0.02	0.01
1966	**0.45**	0.10	0.35	0.12	0.14	0.06	0.02	0.01
1967	**0.45**	0.11	0.34	0.12	0.14	0.06	0.02	0.01
1968	**0.45**	0.11	0.34	0.12	0.14	0.05	0.02	0.01
1969	**0.46**	0.11	0.34	0.12	0.14	0.05	0.02	0.01
1970	**0.47**	0.12	0.35	0.13	0.14	0.05	0.02	0.01
1971	**0.48**	0.12	0.35	0.14	0.14	0.05	0.02	0.01
35 and over								
1920	**0.30**	0.02	0.28	0.03	0.06	0.06	0.05	0.08
1925	**0.30**	0.02	0.28	0.03	0.06	0.06	0.05	0.08
1930	**0.26**	0.02	0.25	0.03	0.05	0.06	0.04	0.07
1935	**0.17**	0.01	0.16	0.02	0.03	0.04	0.03	0.04
1940	**0.12**	0.01	0.11	0.02	0.03	0.03	0.02	0.02
1941	**0.12**	0.01	0.11	0.02	0.03	0.03	0.02	0.02
1942	**0.12**	0.01	0.11	0.02	0.03	0.03	0.02	0.02
1943	**0.12**	0.01	0.11	0.02	0.03	0.03	0.02	0.02
1944	**0.13**	0.01	0.12	0.02	0.03	0.03	0.02	0.02
1945	**0.13**	0.01	0.12	0.02	0.03	0.03	0.02	0.02
1946	**0.14**	0.02	0.12	0.02	0.03	0.03	0.02	0.02
1947	**0.14**	0.02	0.12	0.02	0.03	0.03	0.02	0.01
1948	**0.14**	0.02	0.13	0.02	0.04	0.03	0.02	0.02
1949	**0.15**	0.02	0.13	0.02	0.04	0.03	0.02	0.02
1950	**0.16**	0.02	0.14	0.02	0.04	0.04	0.02	0.02
1951	**0.17**	0.03	0.14	0.03	0.04	0.04	0.02	0.02
1952	**0.18**	0.03	0.15	0.03	0.05	0.04	0.02	0.02
1953	**0.19**	0.03	0.15	0.03	0.05	0.04	0.02	0.02
1954	**0.19**	0.04	0.16	0.03	0.05	0.04	0.02	0.02
1955	**0.20**	0.04	0.16	0.03	0.05	0.04	0.02	0.02
1956	**0.21**	0.04	0.16	0.04	0.05	0.04	0.02	0.02
1957	**0.21**	0.05	0.16	0.04	0.06	0.04	0.02	0.02
1958	**0.22**	0.05	0.17	0.04	0.06	0.04	0.02	0.02
1959	**0.23**	0.06	0.17	0.04	0.06	0.04	0.02	0.01
1960	**0.24**	0.06	0.18	0.04	0.06	0.04	0.02	0.01

Table 10.5 Estimated distribution of women of child-bearing age by number of live-born children (percentages)[1]: year of birth and age, 1920-1985

England and Wales

Year of birth of woman	Age of woman (completed years)	Number of live-born children[2]				
		0 (Childless women)	1	2	3	4 or more
1920	20	89	9	2	0	0
1925		88	11	2	0	0
1930		84	13	3	0	0
1935		83	13	3	0	0
1940		78	16	4	1	0
1945		74	18	6	1	0
1950		74	18	7	1	0
1955		77	16	6	1	0
1960		82	13	4	1	0
1965		85	11	4	1	0
1970		82	13	3	1	0
1975		84	12	3	1	0
1980		83	13	3	1	0
1985		85	11	3	1	0
1920	25	55	28	12	4	1
1925		46	32	16	5	2
1930		44	30	17	6	3
1935		39	30	21	7	3
1940		34	27	25	9	5
1945		34	27	27	9	4
1950		41	24	25	7	2
1955		48	22	22	6	2
1960		55	20	18	6	2
1965		59	19	15	5	1
1970		61	19	14	5	1
1975		65	17	13	4	1
1980		64	18	12	4	2
1920	30	30	28	25	11	6
1925		26	29	27	12	7
1930		22	26	30	14	9
1935		18	21	32	17	12
1940		17	17	35	19	12
1945		16	19	41	17	8
1950		22	19	40	14	6
1955		27	19	35	13	5
1960		33	18	31	12	5
1965		36	20	28	11	5
1970		39	22	24	10	4
1975		44	22	22	9	4
1920	35	23	23	28	15	11
1925		19	24	29	16	12
1930		15	20	31	18	16
1935		13	16	32	21	18
1940		12	14	36	22	16
1945		11	15	43	20	11
1950		16	14	43	18	9
1955		19	15	40	18	9
1960		23	14	37	18	9
1965		24	17	36	16	8
1970		25	19	34	14	8
1920	40	21	22	27	16	14
1925		17	22	28	17	16
1930		14	18	30	19	19
1935		12	15	32	21	20
1940		11	13	36	22	18
1945		10	14	43	21	12
1950		14	13	44	19	10
1955		16	13	41	19	10
1960		19	13	38	20	10
1965		20	15	38	18	9
1920	45[3]	21	21	27	16	15
1925		17	22	28	17	16
1930		13	18	30	19	20
1935		12	15	32	21	20
1940		11	13	36	22	18
1945		10	14	43	21	12
1950		13	13	44	20	10
1955		16	13	41	20	10
1960		18	13	38	20	10

Note: Figures may not add exactly due to rounding - see section 2.15.
1 See sections 2.9 and 3.9.
2 Estimates including births within and outside marriage - see section 2.9.
3 Includes births at ages over 45.

**Table 11.1 Live births within marriage: estimated distribution
by socio-economic classification of father as defined by occupation,
by number of previous live-born children and age of mother, 2001–2005**

England and Wales
thousands

Year	Number of previous live-born children					Year	Number of previous live-born children				
	Total	0	1	2	3 or more		Total	0	1	2	3 or more
Socio-economic classification 1.1						**Socio-economic classification 2**					
All ages of mother at birth						**All ages of mother at birth**					
2001	**34.1**	14.8	13.6	4.4	1.3	2001	**82.5**	36.4	31.0	11.1	4.0
2002	**35.1**	15.0	14.1	4.5	1.5	2002	**81.1**	35.9	30.9	10.0	4.2
2003	**35.7**	15.3	14.3	4.7	1.5	2003	**86.1**	38.6	32.4	10.8	4.3
2004	**35.3**	15.3	13.8	4.9	1.3	2004	**85.5**	38.0	32.5	10.6	4.4
2005	**35.9**	15.9	14.1	4.4	1.4	2005	**86.0**	38.6	32.3	10.8	4.3
Under 20						**Under 20**					
2001	**0.1**	0.0	0.0	0.0	0.0	2001	**0.5**	0.4	0.1	0.0	0.0
2002	**0.0**	0.0	0.0	0.0	0.0	2002	**0.5**	0.5	0.1	0.0	0.0
2003	**0.0**	0.0	0.0	0.0	0.0	2003	**0.4**	0.3	0.0	0.0	0.0
2004	**0.0**	0.0	0.0	0.0	0.0	2004	**0.3**	0.3	0.0	0.0	0.0
2005	**0.0**	0.0	0.0	0.0	0.0	2005	**0.3**	0.3	0.0	0.0	0.0
20-24						**20-24**					
2001	**1.1**	0.8	0.3	0.1	0.0	2001	**6.3**	4.0	1.9	0.4	0.0
2002	**1.1**	0.7	0.4	0.0	0.0	2002	**6.3**	3.9	1.9	0.4	0.1
2003	**1.2**	0.7	0.4	0.0	0.0	2003	**6.6**	4.1	1.9	0.5	0.1
2004	**1.0**	0.6	0.3	0.0	0.0	2004	**6.6**	4.0	2.1	0.4	0.1
2005	**1.2**	0.8	0.3	0.0	0.0	2005	**6.5**	3.8	2.1	0.6	0.1
25-29						**25-29**					
2001	**7.7**	4.5	2.4	0.7	0.1	2001	**22.7**	12.7	7.1	2.0	0.7
2002	**7.1**	4.2	2.2	0.5	0.1	2002	**22.1**	12.6	7.1	1.9	0.6
2003	**7.1**	4.2	2.2	0.5	0.2	2003	**22.1**	12.4	7.2	1.7	0.7
2004	**6.7**	3.9	2.2	0.5	0.1	2004	**21.7**	12.2	6.9	2.1	0.6
2005	**6.8**	4.1	2.0	0.5	0.2	2005	**22.0**	12.7	6.8	1.9	0.6
30 and over						**30 and over**					
2001	**25.1**	9.4	10.8	3.7	1.2	2001	**52.9**	19.1	22.0	8.6	3.3
2002	**26.9**	10.0	11.5	4.0	1.4	2002	**52.2**	19.0	21.8	7.7	3.6
2003	**27.5**	10.4	11.6	4.2	1.3	2003	**57.1**	21.8	23.2	8.6	3.6
2004	**27.6**	10.7	11.3	4.4	1.2	2004	**56.9**	21.5	23.4	8.1	3.8
2005	**28.0**	11.0	11.7	3.9	1.2	2005	**57.3**	21.8	23.5	8.3	3.7
Socio-economic classification 1.2						**Socio-economic classification 3**					
All ages of mother at birth						**All ages of mother at birth**					
2001	**47.3**	21.1	17.7	6.0	2.4	2001	**21.1**	9.5	8.0	2.4	1.0
2002	**47.8**	22.4	17.4	5.8	2.1	2002	**21.4**	10.4	7.5	2.5	1.0
2003	**49.5**	22.9	18.3	6.3	2.0	2003	**22.4**	10.7	8.5	2.4	0.8
2004	**50.0**	23.5	18.0	6.3	2.2	2004	**22.9**	11.5	8.2	2.2	1.0
2005	**49.2**	23.5	17.8	5.7	2.3	2005	**22.9**	11.4	8.0	2.2	1.3
Under 20						**Under 20**					
2001	**0.1**	0.1	0.0	0.0	0.0	2001	**0.4**	0.4	0.1	0.0	0.0
2002	**0.1**	0.1	0.0	0.0	0.0	2002	**0.3**	0.3	0.0	0.0	0.0
2003	**0.1**	0.1	0.0	0.0	0.0	2003	**0.2**	0.2	0.0	0.0	0.0
2004	**0.1**	0.1	0.0	0.0	0.0	2004	**0.4**	0.3	0.1	0.0	0.0
2005	**0.1**	0.1	0.0	0.0	0.0	2005	**0.2**	0.2	0.0	0.0	0.0
20-24						**20-24**					
2001	**2.2**	1.6	0.5	0.1	0.0	2001	**2.7**	1.7	0.8	0.2	0.0
2002	**2.0**	1.4	0.5	0.1	0.0	2002	**2.7**	1.8	0.7	0.1	0.1
2003	**2.2**	1.5	0.6	0.1	0.0	2003	**3.0**	1.8	1.0	0.2	0.0
2004	**2.3**	1.6	0.5	0.1	0.0	2004	**3.0**	1.9	0.9	0.2	0.0
2005	**1.9**	1.3	0.5	0.1	0.0	2005	**2.6**	1.8	0.7	0.1	0.0
25.9						**25-29**					
2001	**10.6**	6.4	3.3	0.6	0.3	2001	**6.2**	3.1	2.3	0.5	0.2
2002	**10.4**	6.5	3.0	0.7	0.2	2002	**6.5**	3.6	2.0	0.7	0.2
2003	**10.9**	7.0	3.1	0.6	0.2	2003	**6.8**	3.8	2.2	0.6	0.2
2004	**10.7**	6.7	3.1	0.8	0.1	2004	**7.0**	3.9	2.2	0.6	0.2
2005	**10.7**	6.7	3.0	0.6	0.3	2005	**7.1**	4.0	2.2	0.5	0.3

Note: For a description of socio-economic classifications and the sample used in calculation - see section 3.10.

Table 11.1 - *continued*

England and Wales
thousands

Year	Number of previous live-born children					Year	Number of previous live-born children				
	Total	0	1	2	3 or more		**Total**	0	1	2	3 or more
	30 and over						**30 and over**				
2001	**34.2**	12.9	13.9	5.3	2.1	2001	**11.8**	4.3	4.9	1.8	0.8
2002	**35.2**	14.3	13.8	5.1	1.9	2002	**11.9**	4.7	4.8	1.7	0.7
2003	**36.4**	14.3	14.7	5.6	1.8	2003	**12.3**	4.9	5.2	1.6	0.7
2004	**37.0**	15.1	14.5	5.4	2.1	2004	**12.5**	5.3	5.1	1.4	0.7
2005	**36.6**	15.4	14.2	5.0	2.0	2005	**13.0**	5.3	5.1	1.6	1.0
	Socio-economic classification 4						**Socio-economic classification 6**				
	All ages of mother at birth						**All ages of mother at birth**				
2001	**41.9**	14.1	15.5	7.2	5.1	2001	**33.6**	12.6	11.8	5.6	3.6
2002	**43.8**	14.7	15.9	8.2	5.1	2002	**34.0**	13.0	11.6	5.4	4.0
2003	**44.8**	15.5	16.1	8.2	5.0	2003	**31.8**	11.7	11.1	5.1	3.9
2004	**47.8**	16.8	17.3	8.3	5.5	2004	**34.9**	13.8	11.3	5.8	4.0
2005	**48.0**	17.2	16.5	8.5	5.8	2005	**34.8**	13.4	11.6	5.7	4.1
	Under 20						**Under 20**				
2001	**0.6**	0.5	0.1	0.0	0.0	2001	**0.9**	0.8	0.1	0.0	0.0
2002	**0.5**	0.4	0.1	0.0	0.0	2002	**1.0**	0.8	0.2	0.0	0.0
2003	**0.5**	0.4	0.1	0.0	0.0	2003	**0.9**	0.7	0.2	0.0	0.0
2004	**0.5**	0.4	0.0	0.0	0.0	2004	**0.8**	0.6	0.2	0.0	0.0
2005	**0.6**	0.5	0.1	0.0	0.0	2005	**0.6**	0.5	0.1	0.0	0.0
	20-24						**20-24**				
2001	**4.7**	2.4	1.7	0.6	0.1	2001	**7.8**	3.7	2.9	0.9	0.3
2002	**4.8**	2.3	1.8	0.6	0.1	2002	**7.8**	4.1	2.7	0.9	0.2
2003	**5.1**	2.6	1.7	0.6	0.2	2003	**6.8**	3.3	2.6	0.7	0.2
2004	**5.2**	2.6	2.0	0.5	0.1	2004	**7.8**	4.0	2.6	1.0	0.2
2005	**5.0**	2.6	1.8	0.6	0.1	2005	**6.9**	3.7	2.3	0.7	0.2
	25-29						**25-29**				
2001	**12.0**	4.7	4.7	1.8	0.9	2001	**11.2**	4.2	4.0	2.1	0.9
2002	**11.6**	4.5	4.1	2.0	0.9	2002	**10.7**	4.1	3.7	1.8	1.1
2003	**12.3**	4.8	4.4	2.2	0.9	2003	**9.9**	3.8	3.3	1.7	1.1
2004	**13.0**	5.5	4.3	2.0	1.1	2004	**10.6**	4.5	3.5	1.7	1.0
2005	**13.4**	5.7	4.4	2.2	1.1	2005	**11.1**	4.3	3.9	1.9	1.0
	30 and over						**30 and over**				
2001	**24.6**	6.6	9.1	4.8	4.1	2001	**13.8**	4.0	4.8	2.7	2.4
2002	**26.9**	7.5	9.8	5.6	4.0	2002	**14.5**	4.0	5.1	2.6	2.7
2003	**26.9**	7.7	9.9	5.4	3.9	2003	**14.0**	3.9	4.9	2.6	2.6
2004	**29.1**	8.2	11.0	5.8	4.2	2004	**15.7**	4.8	5.0	3.1	2.8
2005	**29.0**	8.5	10.1	5.8	4.6	2005	**16.1**	5.0	5.2	3.0	2.9
	Socio-economic classification 5						**Socio-economic classification 7**				
	All ages of mother at birth						**All ages of mother at birth**				
2001	**45.2**	17.4	17.8	6.5	3.5	2001	**38.4**	13.4	13.2	6.5	5.3
2002	**43.3**	17.2	16.9	5.9	3.3	2002	**33.2**	11.2	11.9	5.6	4.5
2003	**41.9**	16.8	15.9	5.8	3.3	2003	**34.9**	13.0	11.5	5.9	4.4
2004	**42.6**	17.0	16.0	6.1	3.5	2004	**33.4**	11.8	11.8	5.6	4.2
2005	**40.4**	16.6	14.9	5.8	3.0	2005	**34.0**	12.3	11.7	6.0	4.0
	Under 20						**Under 20**				
2001	**0.5**	0.4	0.1	0.0	0.0	2001	**1.0**	0.8	0.2	0.0	0.0
2002	**0.4**	0.4	0.0	0.0	0.0	2002	**1.0**	0.8	0.2	0.0	0.0
2003	**0.5**	0.4	0.1	0.0	0.0	2003	**0.9**	0.8	0.2	0.0	0.0
2004	**0.4**	0.3	0.1	0.0	0.0	2004	**0.8**	0.6	0.1	0.0	0.0
2005	**0.3**	0.3	0.0	0.0	0.0	2005	**0.6**	0.5	0.1	0.0	0.0
	20-24						**20-24**				
2001	**5.6**	3.0	2.0	0.4	0.1	2001	**7.3**	3.4	2.7	0.9	0.2
2002	**5.6**	2.9	2.0	0.6	0.1	2002	**7.2**	3.5	2.7	0.9	0.2
2003	**5.6**	2.8	2.1	0.6	0.1	2003	**6.8**	3.3	2.5	0.7	0.3
2004	**5.0**	2.5	1.8	0.5	0.1	2004	**6.8**	3.3	2.4	0.9	0.2
2005	**4.9**	2.5	1.8	0.5	0.1	2005	**7.0**	3.4	2.5	0.8	0.2

Note: For a description of socio-economic classifications and the sample used in calculation - see section 3.10.

Table 11.1 - *continued*

Left panel

Year	Number of previous live-born children				
	Total	0	1	2	3 or more

25-29

Year	Total	0	1	2	3 or more
2001	**15.6**	6.8	6.0	1.9	0.9
2002	**14.0**	6.2	5.5	1.5	0.7
2003	**12.8**	5.9	4.5	1.8	0.6
2004	**13.1**	6.2	4.5	1.6	0.8
2005	**12.7**	6.1	4.3	1.6	0.7

30 and over

Year	Total	0	1	2	3 or more
2001	**23.5**	7.2	9.6	4.2	2.5
2002	**23.3**	7.7	9.4	3.8	2.4
2003	**22.9**	7.7	9.2	3.4	2.7
2004	**24.2**	8.0	9.7	3.9	2.6
2005	**22.4**	7.7	8.8	3.7	2.2

Socio-economic classification 8

All ages of mother at birth

Year	Total	0	1	2	3 or more
2001	**0.1**	0.0	0.0	0.0	0.0
2002	**0.1**	0.0	0.0	0.0	0.0
2003	**0.0**	0.0	0.0	0.0	0.0
2004	**0.1**	0.0	0.0	0.0	0.0
2005	**0.0**	0.0	0.0	0.0	0.0

Under 20

Year	Total	0	1	2	3 or more
2001	**0.0**	0.0	0.0	0.0	0.0
2002	**0.0**	0.0	0.0	0.0	0.0
2003	**0.0**	0.0	0.0	0.0	0.0
2004	**0.0**	0.0	0.0	0.0	0.0
2005	**0.0**	0.0	0.0	0.0	0.0

20-24

Year	Total	0	1	2	3 or more
2001	**0.0**	0.0	0.0	0.0	0.0
2002	**0.0**	0.0	0.0	0.0	0.0
2003	**0.0**	0.0	0.0	0.0	0.0
2004	**0.0**	0.0	0.0	0.0	0.0
2005	**0.0**	0.0	0.0	0.0	0.0

25-29

Year	Total	0	1	2	3 or more
2001	**0.0**	0.0	0.0	0.0	0.0
2002	**0.0**	0.0	0.0	0.0	0.0
2003	**0.0**	0.0	0.0	0.0	0.0
2004	**0.0**	0.0	0.0	0.0	0.0
2005	**0.0**	0.0	0.0	0.0	0.0

30 and over

Year	Total	0	1	2	3 or more
2001	**0.0**	0.0	0.0	0.0	0.0
2002	**0.0**	0.0	0.0	0.0	0.0
2003	**0.0**	0.0	0.0	0.0	0.0
2004	**0.0**	0.0	0.0	0.0	0.0
2005	**0.0**	0.0	0.0	0.0	0.0

Right panel

Year	Number of previous live-born children				
	Total	0	1	2	3 or more

25-29

Year	Total	0	1	2	3 or more
2001	**13.1**	4.9	4.6	2.4	1.3
2002	**10.5**	3.6	3.8	1.9	1.2
2003	**11.4**	4.4	3.8	2.2	1.1
2004	**10.5**	3.8	3.8	1.9	1.1
2005	**11.0**	4.2	3.8	2.1	0.9

30 and over

Year	Total	0	1	2	3 or more
2001	**17.1**	4.4	5.7	3.3	3.7
2002	**14.5**	3.3	5.3	2.8	3.1
2003	**15.7**	4.5	5.0	3.0	3.1
2004	**15.2**	4.2	5.3	2.8	2.9
2005	**15.2**	4.2	5.2	3.0	2.9

All socio-economic classifications (including 'not classified')

All ages of mother at birth

Year	Total	0	1	2	3 or more
2001	**356.5**	143.9	132.2	52.1	28.3
2002	**354.1**	145.2	130.3	50.3	28.2
2003	**364.2**	151.0	132.9	52.0	28.4
2004	**370.0**	154.5	133.7	52.5	29.3
2005	**369.3**	156.0	132.0	52.2	29.2

Under 20

Year	Total	0	1	2	3 or more
2001	**4.6**	3.8	0.8	0.1	0.0
2002	**4.6**	3.8	0.7	0.1	0.0
2003	**4.3**	3.5	0.8	0.1	0.0
2004	**4.1**	3.3	0.7	0.1	0.0
2005	**3.7**	3.0	0.6	0.1	0.0

20-24

Year	Total	0	1	2	3 or more
2001	**40.7**	22.2	13.7	3.9	0.9
2002	**40.7**	22.4	13.5	3.9	0.9
2003	**40.9**	22.2	13.9	3.8	1.0
2004	**41.3**	22.6	13.8	4.0	0.9
2005	**40.0**	22.1	13.2	3.8	0.9

25-29

Year	Total	0	1	2	3 or more
2001	**103.1**	48.8	35.6	12.8	5.9
2002	**97.6**	47.1	33.0	11.8	5.6
2003	**98.7**	48.4	32.5	12.1	5.7
2004	**98.5**	48.9	31.9	12.1	5.7
2005	**100.0**	50.0	32.1	12.3	5.6

30 and over

Year	Total	0	1	2	3 or more
2001	**208.0**	69.1	82.1	35.4	21.5
2002	**211.2**	71.9	83.2	34.5	21.6
2003	**220.3**	76.9	85.7	36.0	21.8
2004	**226.1**	79.8	87.3	36.4	22.7
2005	**225.7**	80.9	86.1	36.1	22.7

Table 11.2 Pre-maritally conceived first live births to married women (numbers and percentages): estimated distribution by socio-economic classification of father as defined by occupation, and age of mother, 2001-2005

England and Wales

Year	All socio-economic classifications (including 'not classified')	Socio-economic classification of father								
		1.1	1.2	2	3	4	5	6	7	8
Number (thousands)										
All ages of mother at birth										
2001	**13.6**	1.1	1.4	3.4	1.1	1.2	1.8	1.2	1.6	0.0
2002	**13.9**	1.0	1.5	3.0	1.1	1.7	1.8	1.6	1.4	0.0
2003	**14.5**	1.2	1.5	2.9	1.0	1.7	1.9	1.4	1.9	0.0
2004	**14.5**	1.0	1.5	3.2	1.3	1.7	1.7	1.7	1.4	0.0
2005	**13.6**	0.9	1.5	3.1	0.9	1.8	1.6	1.3	1.5	0.0
Under 20										
2001	**1.1**	0.0	0.0	0.1	0.1	0.1	0.1	0.2	0.2	0.0
2002	**1.1**	0.0	0.0	0.1	0.1	0.1	0.1	0.2	0.3	0.0
2003	**1.0**	0.0	0.0	0.1	0.1	0.1	0.1	0.2	0.2	0.0
2004	**1.0**	0.0	0.0	0.1	0.1	0.1	0.1	0.2	0.2	0.0
2005	**0.9**	0.0	0.0	0.1	0.1	0.1	0.1	0.1	0.2	0.0
20-24										
2001	**3.8**	0.2	0.2	0.9	0.4	0.3	0.6	0.4	0.5	0.0
2002	**3.6**	0.1	0.1	0.7	0.4	0.4	0.5	0.6	0.6	0.0
2003	**3.8**	0.1	0.1	0.6	0.4	0.4	0.7	0.4	0.7	0.0
2004	**3.9**	0.1	0.2	0.7	0.4	0.4	0.5	0.6	0.5	0.0
2005	**3.5**	0.1	0.2	0.6	0.3	0.5	0.4	0.5	0.5	0.0
25-29										
2001	**4.1**	0.3	0.5	1.0	0.3	0.4	0.5	0.4	0.5	0.0
2002	**4.1**	0.4	0.5	0.9	0.2	0.5	0.7	0.4	0.3	0.0
2003	**4.1**	0.3	0.5	1.0	0.2	0.4	0.5	0.4	0.5	0.0
2004	**4.2**	0.3	0.5	0.9	0.4	0.6	0.5	0.5	0.3	0.0
2005	**4.0**	0.3	0.4	1.0	0.3	0.5	0.5	0.4	0.5	0.0
30 and over										
2001	**4.6**	0.5	0.7	1.4	0.2	0.4	0.6	0.3	0.4	0.0
2002	**5.1**	0.5	0.8	1.3	0.4	0.7	0.5	0.4	0.3	0.0
2003	**5.6**	0.7	0.9	1.2	0.3	0.7	0.6	0.4	0.5	0.0
2004	**5.5**	0.6	0.8	1.5	0.3	0.6	0.6	0.4	0.4	0.0
2005	**5.3**	0.5	0.9	1.4	0.3	0.7	0.5	0.3	0.4	0.0
As a percentage of all first live births within marriage[1]										
All ages of mother at birth										
2001	**9.4**	7.2	6.8	9.3	11.4	8.8	10.4	9.9	11.9	0.0
2002	**9.6**	6.9	6.6	8.4	10.8	11.3	10.2	12.4	12.7	0.0
2003	**9.6**	7.6	6.7	7.5	9.0	10.6	11.2	11.7	14.8	-
2004	**9.4**	6.5	6.3	8.4	11.1	10.2	9.8	11.9	11.9	-
2005	**8.7**	5.8	6.4	7.9	8.1	10.2	9.6	9.8	12.2	*
Under 20										
2001	**28.7**	67.4	33.7	30.8	39.3	22.5	30.0	30.4	27.0	-
2002	**29.3**	48.9	30.1	28.8	38.5	24.5	22.0	28.8	33.0	0.0
2003	**29.8**	0.0	39.5	21.1	42.2	33.6	26.5	23.4	31.8	-
2004	**30.3**	113.3	16.2	27.3	46.0	16.2	40.2	25.6	36.1	-
2005	**29.3**	0.0	51.4	27.4	33.6	16.8	36.3	24.1	35.4	*

Note: For a description of socio-economic classification and the sample used in calculation - see section 3.10.

1 Some cell counts have been suppressed to protect the confidentiality of individuals - see section 2.15.

Table 11.2 - *continued*

Year	All socio-economic classifications (including 'not classified')	Socio-economic classification of father								
		1.1	1.2	2	3	4	5	6	7	8
	20-24									
2001	**16.9**	25.9	11.6	21.7	23.3	14.0	19.0	11.1	15.8	0.0
2002	**16.1**	20.3	7.6	18.0	21.2	16.8	17.0	13.6	16.2	0.0
2003	**17.2**	20.5	9.7	15.7	19.4	15.9	22.9	11.9	19.9	-
2004	**17.1**	12.2	11.4	17.0	22.1	15.6	18.0	15.9	16.2	-
2005	**15.7**	15.1	13.6	16.3	15.7	18.1	16.6	13.2	13.5	*
	25-29									
2001	**8.4**	6.5	7.6	7.7	9.7	8.0	7.9	8.5	10.4	0.0
2002	**8.6**	8.4	8.0	7.0	5.6	10.0	11.1	10.3	8.8	-
2003	**8.4**	7.7	6.4	8.0	5.8	7.6	8.8	11.4	11.1	-
2004	**8.6**	6.6	6.9	7.4	9.6	10.5	8.4	11.1	7.7	-
2005	**8.1**	6.4	5.5	7.5	6.5	9.4	8.8	9.2	10.9	*
	30 and over									
2001	**6.7**	5.6	5.5	7.2	5.7	6.5	8.2	6.5	8.1	-
2002	**7.1**	5.2	5.7	6.9	9.1	9.7	6.3	9.9	8.4	-
2003	**7.2**	6.7	6.3	5.5	6.6	9.7	8.1	9.4	11.8	-
2004	**6.9**	5.7	5.3	7.0	6.4	7.8	7.2	8.1	9.4	-
2005	**6.5**	4.9	5.9	6.5	5.6	8.1	6.9	6.5	9.8	*

Table 11.3 Median birth intervals: socio-economic classification of father as defined
by occupation (for interval to first birth), and mother's marriage order, 2001-2005

United Kingdom[2]
England and Wales

Year	Median intervals in months[1]										First to second birth	Second to third birth	Third to fourth birth
	Marriage to first birth												
	All socio-economic classifications (including 'not classified') 1.1	1.2	2	3	4	5	6	7	8				
	Women married once only (England and Wales)										All women (United Kingdom)[1,2]		
2001	**27**	32	30	28	25	25	26	25	23	12	36	41	38
2002	**26**	32	31	28	25	24	27	23	21	12	37	40	37
2003	**26**	30	30	27	25	24	25	23	24	-	37	42	38
2004	**26**	31	30	27	25	24	25	22	22	-	37	41	36
2005	**25**	29	29	26	25	24	25	24	23	3	36	40	35
	Remarried women (England and Wales)												
2001	**18**	19	19	18	19	20	17	16	22	-			
2002	**17**	18	16	19	21	17	16	14	22	-			
2003	**18**	19	19	20	20	16	17	18	15	-			
2004	**18**	18	19	17	21	16	21	16	15	-			
2005	**19**	23	17	20	19	18	19	14	12	-			

Note: For a description of socio-economic classifications and the sample used in calculation - see section 3.10.
1 See section 3.11.
2 Incorrectly labelled 'Great Britain' in the 2002 volume.

Table 11.4 Mean age of women at live births within marriage: socio-economic
classification of father as defined by occupation, and birth order[1], 2001-2005

England and Wales

Year	All socio-economic classifications (including 'not classified')	Socio-economic classification of father								
		1.1	1.2	2	3	4	5	6	7	8
	All live births within marriage									
2001	**30.9**	32.6	32.5	31.7	30.5	31.1	30.4	28.9	29.4	26.5
2002	**31.0**	32.8	32.7	31.6	30.5	31.2	30.4	29.0	29.2	25.1
2003	**31.2**	32.9	32.6	31.8	30.5	31.2	30.5	29.2	29.3	33.2
2004	**31.2**	33.1	32.6	31.8	30.5	31.3	30.7	29.3	29.5	28.3
2005	**31.3**	33.0	32.8	31.9	30.9	31.3	30.7	29.5	29.5	20.5
	First live births within marriage									
2001	**29.7**	31.5	31.1	30.4	29.2	29.7	29.3	27.5	27.8	26.3
2002	**29.7**	31.7	31.5	30.3	29.2	29.9	29.3	27.3	27.1	20.5
2003	**29.9**	31.7	31.3	30.6	29.4	29.9	29.4	27.6	27.9	-
2004	**30.0**	32.0	31.4	30.8	29.5	29.9	29.6	27.9	28.1	-
2005	**30.1**	31.9	31.7	30.7	29.7	29.9	29.6	28.2	28.0	21.5
	Second live births within marriage									
2001	**31.2**	33.1	33.0	32.2	31.0	30.9	30.4	28.8	29.2	26.8
2002	**31.4**	33.2	33.2	32.2	31.2	31.2	30.6	29.2	29.3	25.5
2003	**31.5**	33.3	33.2	32.3	31.1	31.3	30.6	29.1	29.2	26.5
2004	**31.6**	33.3	33.1	32.2	31.1	31.4	30.9	29.2	29.4	24.5
2005	**31.7**	33.5	33.2	32.5	31.6	31.3	31.0	29.3	29.4	18.5
	Third live births within marriage									
2001	**32.4**	34.2	34.8	33.6	32.7	32.2	32.1	30.2	30.4	-
2002	**32.3**	34.6	34.5	33.1	32.6	32.3	31.7	30.2	30.4	29.0
2003	**32.5**	34.7	34.9	33.4	31.9	32.2	31.3	30.5	30.3	34.5
2004	**32.5**	34.8	34.4	33.3	31.8	32.5	31.8	30.4	30.4	-
2005	**32.5**	34.7	34.6	33.4	32.5	32.3	31.9	30.5	30.3	-

Notes: For a description of socio-economic classifications and the sample used in calculation - see section 3.10.
The mean ages shown in this table are unstandardised and therefore take no account of the structure of the population by age, marital status or parity.
1 See section 2.9.

Table 11.5   Jointly registered live births outside marriage: socio-economic   **England and Wales**
            classification of father as defined by occupation, and age of mother, 2001-2005   *thousands*

Year	Age of mother	All socio-economic classifications (including 'not classified')	Socio-economic classification of father								
			1.1	1.2	2	3	4	5	6	7	8
2001	**All ages**	**194.3**	**7.3**	**9.4**	**29.6**	**9.3**	**25.6**	**33.5**	**28.3**	**40.7**	**0.0**
2002		**198.9**	**7.6**	**9.3**	**30.0**	**9.6**	**28.1**	**33.8**	**28.4**	**40.2**	**0.1**
2003		**212.3**	**8.3**	**10.3**	**32.2**	**11.0**	**30.4**	**34.6**	**28.8**	**42.6**	**0.1**
2004		**224.4**	**8.6**	**10.6**	**33.5**	**11.4**	**33.3**	**37.3**	**30.7**	**43.4**	**0.0**
2005		**231.3**	**8.3**	**10.7**	**35.2**	**11.5**	**33.9**	**38.0**	**31.2**	**44.7**	**0.1**
2001	Under 20	**27.9**	0.2	0.3	1.9	1.1	2.1	4.6	5.6	8.3	0.0
2002		**27.7**	0.2	0.4	1.9	1.0	2.2	3.9	5.6	8.4	0.0
2003		**28.7**	0.2	0.4	1.9	1.2	2.3	4.3	5.4	8.4	0.0
2004		**29.9**	0.2	0.3	1.7	1.4	2.8	4.4	5.4	8.5	0.0
2005		**30.4**	0.1	0.3	2.1	1.2	2.5	4.3	5.7	8.4	0.0
2001	20-24	**54.5**	0.9	1.4	6.8	2.5	6.3	10.2	9.3	13.7	0.0
2002		**56.6**	1.1	1.4	6.3	2.8	6.8	11.3	9.8	13.5	0.0
2003		**61.3**	1.0	1.4	7.3	3.2	7.7	10.6	10.0	15.5	0.0
2004		**65.4**	1.1	1.4	7.7	3.3	8.6	12.0	11.0	15.4	0.0
2005		**67.7**	1.1	1.4	8.1	3.5	8.7	12.6	10.6	16.1	0.0
2001	25-29	**48.2**	1.9	2.4	7.7	2.6	6.8	8.8	6.9	9.4	0.0
2002		**47.7**	1.9	2.3	7.9	2.5	7.3	8.5	6.3	9.0	0.0
2003		**49.7**	2.1	2.3	8.1	2.7	7.6	8.9	6.4	9.1	0.0
2004		**52.7**	1.8	2.4	8.8	2.7	8.3	9.5	7.3	9.2	0.0
2005		**55.5**	2.0	2.5	8.7	2.6	9.2	9.6	7.7	9.9	0.0
2001	30 and over	**63.7**	4.3	5.3	13.2	3.2	10.4	10.0	6.5	9.2	0.0
2002		**66.9**	4.4	5.2	13.8	3.3	11.9	10.2	6.7	9.4	0.0
2003		**72.7**	5.1	6.3	14.9	3.9	12.8	10.8	6.9	9.6	0.0
2004		**76.4**	5.4	6.5	15.4	4.0	13.6	11.4	7.0	10.3	0.0
2005		**77.8**	5.1	6.5	16.4	4.2	13.5	11.5	7.1	10.2	0.0

Note: For a description of socio-economic classifications and the sample used in calculation - see section 3.10.

Appendix Table 1 Estimated resident population[1]: sex and age, 1995-2005 England and Wales

thousands

	Age	Year										
		1995	1996	1997	1998	1999	2000	2001	2002	2003	2004	2005
Persons	**All ages**	**51,272.0**	**51,410.4**	**51,559.6**	**51,720.1**	**51,933.5**	**52,140.2**	**52,360.0**	**52,570.2**	**52,793.7**	**53,046.2**	**53,390.2**
Males	All ages	24,946.3	25,029.8	25,113.2	25,200.7	25,323.4	25,438.0	25,574.3	25,702.4	25,840.6	25,988.2	26,178.8
Females	All ages	26,325.6	26,380.7	26,446.5	26,519.4	26,610.0	26,702.1	26,785.7	26,867.8	26,953.1	27,058.0	27,211.5
	15-44	**10,711.9**	**10,717.6**	**10,727.0**	**10,737.2**	**10,766.6**	**10,820.8**	**10,869.5**	**10,905.2**	**10,944.1**	**10,983.7**	**11,057.5**
	15-19	1,473.8	1,501.4	1,534.3	1,560.7	1,566.5	1,562.7	1,578.4	1,609.3	1,649.6	1,679.0	1,702.9
	20-24	1,712.0	1,633.0	1,560.6	1,516.6	1,516.9	1,540.0	1,577.7	1,604.4	1,636.8	1,665.3	1,702.7
	25-29	2,004.9	1,979.7	1,945.2	1,902.3	1,850.9	1,809.9	1,744.7	1,674.0	1,627.5	1,625.4	1,663.4
	30-34	2,051.0	2,075.8	2,087.9	2,080.7	2,069.2	2,049.3	2,033.4	2,010.3	1,974.5	1,916.4	1,864.4
	35-39	1,802.5	1,853.9	1,903.4	1,951.9	2,004.0	2,053.7	2,082.0	2,102.5	2,101.0	2,092.5	2,070.6
	40-44	1,667.7	1,673.8	1,695.6	1,725.0	1,759.1	1,805.2	1,853.3	1,904.8	1,954.6	2,005.1	2,053.6
	45-49	1,821.8	1,825.4	1,744.4	1,693.2	1,673.9	1,664.6	1,670.1	1689.7	1,719.9	1,750.3	1,792.7

Note: Figures may not add exactly due to rounding - see section 2.15.
1 See section 2.1.

Appendix Table 2 Estimated resident female population[1]: age and marital status, 1995–2005 England and Wales

thousands

	Age	Year										
		1995	1996	1997	1998	1999	2000	2001	2002	2003	2004	2005
Married	**15-44**	**5,178.1**	**5,070.5**	**4,962.6**	**4,859.5**	**4,772.1**	**4,696.7**	**4,611.1**	**4,490.8**	**4,372.4**	**4,262.0**	**4,157.0**
	15-19	20.8	20.7	20.3	20.3	19.8	17.6	16.2	13.0	11.7	10.6	8.9
	20-24	300.0	259.6	224.9	201.1	187.8	179.6	178.0	166.1	160.8	156.4	148.7
	25-29	962.7	906.3	844.3	725.1	677.2	625.4	567.4	523.5	496.6	483.3	
	30-34	1,330.4	1,315.8	1,293.1	1,259.4	1,222.8	1,182.2	1,142.4	1,094.5	1,042.6	984.6	931.8
	35-39	1,305.5	1,319.9	1,331.5	1,342.0	1,354.1	1,362.1	1,356.3	1,342.4	1,314.6	1,284.4	1,247.4
	40-44	1,258.7	1,248.1	1,248.4	1,253.9	1,262.5	1,278.0	1,292.9	1,307.4	1,319.1	1,329.4	1,336.9
	45-49	1,413.5	1,402.1	1,323.0	1,268.4	1,239.3	1,218.7	1,208.1	1,206.0	1,211.8	1,216.0	1,227.1
Single, widowed and divorced	**15-44**	**5,533.9**	**5,647.1**	**5,764.4**	**5,877.7**	**5,994.6**	**6,124.1**	**6,258.4**	**6,414.4**	**6,571.7**	**6,721.7**	**6,900.6**
	15-19	1,453.1	1,480.7	1,514.0	1,540.4	1,546.8	1,545.1	1,562.3	1,596.3	1,637.9	1,668.4	1,694.0
	20-24	1,412.0	1,373.3	1,335.7	1,315.6	1,329.1	1,360.5	1,399.7	1,438.3	1,476.0	1,508.9	1,554.0
	25-29	1,042.2	1,073.5	1,100.8	1,119.4	1,125.8	1,132.7	1,119.3	1,106.6	1,104.0	1,128.8	1,180.1
	30-34	720.6	760.0	794.7	821.3	846.4	867.1	891.0	915.8	931.8	931.8	932.7
	35-39	497.0	534.0	571.9	609.9	649.8	691.6	725.7	760.0	786.5	808.1	823.2
	40-44	409.0	425.7	447.3	471.1	496.7	527.3	560.5	597.4	635.5	675.7	716.6
	45-49	408.3	423.3	421.4	424.8	434.6	446.0	462.0	483.7	508.1	534.3	565.5

Note: Figures may not add exactly due to rounding - see section 2.15.
1 See section 2.1.

Appendix Table 3 **Estimated standard errors for**
numbers in analyses by socio-economic classification **England and Wales**

Estimated number[1] (thousands)	Standard error (thousands)	Standard error as a percentage of estimated number
0.5	0.07	14.0
1.0	0.09	9.0
5.0	0.21	4.2
10.0	0.30	3.0
20.0	0.42	2.1
30.0	0.50	1.7
40.0	0.58	1.4
50.0	0.64	1.3
60.0	0.69	1.2
70.0	0.74	1.1
80.0	0.78	1.0
90.0	0.82	0.9
100.0	0.86	0.9
150.0	0.99	0.7
200.0	1.07	0.5
250.0	1.10	0.4
300.0	1.10	0.4

1 See section 3.10 for further details.

Appendix Table 4 **Estimated standard errors for**
percentages in analyses by socio-economic classification **England and Wales**

Estimated number[1] on which percentage is based (thousands)	Percentages					
	5 or 95	10 or 90	20 or 80	30 or 70	40 or 60	50
0.5	2.9	4.0	5.4	6.1	6.6	6.7
1.0	2.1	2.8	3.8	4.3	4.6	4.7
5.0	0.9	1.3	1.7	1.9	2.1	2.1
10.0	0.7	0.9	1.2	1.4	1.5	1.5
20.0	0.5	0.6	0.8	1.0	1.0	1.1
30.0	0.4	0.5	0.7	0.8	0.8	0.9
40.0	0.3	0.4	0.6	0.7	0.7	0.8
50.0	0.3	0.4	0.5	0.6	0.7	0.7
60.0	0.3	0.4	0.5	0.6	0.6	0.6
70.0	0.2	0.3	0.5	0.5	0.6	0.6
80.0	0.2	0.3	0.4	0.5	0.5	0.5
90.0	0.2	0.3	0.4	0.5	0.5	0.5
100.0	0.2	0.3	0.4	0.4	0.5	0.5
150.0	0.2	0.2	0.3	0.4	0.4	0.4
200.0	0.1	0.2	0.3	0.3	0.3	0.3
250.0	0.1	0.2	0.2	0.3	0.3	0.3
300.0	0.1	0.2	0.2	0.3	0.3	0.3

1 See section 3.10 for further details.

Draft entry form used currently for registering live births (Form 309(Rev))

BIRTH

Registration district

Sub-district

Register No.

District & S. Dist. Nos.

Administrative area

Entry No.

CHILD

1. Date and place of birth

2. Name and surname

3. Sex

FATHER

4. Name and surname

5. Place of birth

6. Occupation

MOTHER

7. Name and surname

8. (a) Place of birth

8. (b) Occupation

9. (a) Maiden surname

9. (b) Surname at marriage if different from maiden surname

INFORMANT

10. Usual address (if different from place of child's birth)

11. Name and surname (if not the mother or father)

12. Qualification

13. Usual address (if different from that in 10 above)

14. I certify that the particulars entered above are true to the best of my knowledge and belief

Signature of informant

15. Date of registration

16. Signature of registrar

NHS Number

QA18/5 4/94

FORM 309 (Rev)

BIRTH

GRO Reference No.

District & SD Nos.

Register No.

Date of registration

Entry No.

CONFIDENTIAL PARTICULARS

The particulars below, required under the Population (Statistics) Acts, will not be entered in the register. This confidential information will be used only for the preparation and supply of statistical information by the Registrar General

1. Where the father's name is entered in register: Father's date of birth

   DAY MONTH YEAR

2. In all cases: Mother's date of birth

   DAY MONTH YEAR

3. Where the child is born within marriage:
   (i) Date of marriage

   MONTH YEAR

   (ii) Has the mother been married more than once? YES NO

   (iii) Mother's previous children (excluding birth of births now being registered) by her present husband and any former husband
   (a) Number born alive (including any who have died)
   (b) Number still-born

X is this birth one of twins, triplets, etc* YES NO

If YES, complete (a) and (b)

(a) Total number of live and still-births at this maternity*

   2 3 4 5 6 7+

   Live births
   (b) Entry No. of births

   Still-births
   (b) Entry No. of births

Z

L grams F* title

(i) (ii)

(iii) (iv)

G(a) Father

(va)

H(a)* 1 2 3 4 5

G(b) Mother

(vb)

H(b)* 1 2 3 4 5

POSTCODE

Edit Control

(vi) (vii)

*Tick as appropriate

FORM 309 (Rev)

Draft entry form used currently for registering stillbirths (Form 308(Rev))

STILL-BIRTH

Registration district	Register No.	
Sub-district	District& S. Dist. Nos.	Administrative area

CHILD

1.(a) Date and place of birth

3. Sex

1.(b) Name and surname

2. Cause of death and nature of evidence that child was still-born

a

b

c

d

e

Certified by

FATHER

4. Name and surname

5. Place of birth

6. Occupation

MOTHER

7. Name and surname

8.(a) Place of birth

8.(b) Occupation

9.(a) Maiden surname

9.(b) Surname at marriage if different from maiden surname

10. Usual address (if different from place of child's birth)

INFORMANT

11. Name and surname (if not the mother or father)

12. Qualification

13. Usual address (if different from that in 10 above)

14. I certify that the particulars entered above are true to the best of my knowledge and belief

....
Signature of informant

15. Date of registration

16. Signature of registrar

OA18/3 5/96

FORM 308

STILL-BIRTH

District& SD Nos.	Register No.	
Date of registration	Entry No.	

(i)

(ii)

(iii)

L .... grams   K .... weeks

N Post Mortem Held?* Yes ☐ No ☐

(iv) ME   2   4   6

U

Z

Y*   Before Labour a   During Labour b   Not Known c

(v) a b c d e

M

(vi)

(vii) (viii) (ixa) (ixb)

CONFIDENTIAL PARTICULARS

The particulars below, required under the Population (Statistics) Acts, will not be entered in the register. This confidential information will be used only for the preparation and supply of statistical information by the Registrar General

1. Where the father's name is entered in register:
   Father's date of birth   DAY MONTH YEAR

2. In all cases:
   Mother's date of birth   DAY MONTH YEAR

3. Where the child is born within marriage:
   (i) Date of marriage   MONTH YEAR
   (ii) Has the mother been married more than once?   Yes ☐ No ☐
   (iii) Mother's previous children (excluding birth or births now being registered) by her present husband and any former husband
      (a) Number born alive (including any who have died)
      (b) Number still-born

X Is this birth one of twins, triplets, etc*   YES ☐ NO ☐
   IF YES, complete (a) and (b)

(a) Total number of live and still-births at this maternity*
   2 3 4 5 6 7+

	Live births	Still-births
(b)	Entry No. of births	Entry No. of births

(x)

G(a) Father

H(a)* 1 2 3 4 5

G(b) Mother

H(b)* 1 2 3 4 5

POSTCODE

Edit Control

*Tick as appropriate

FORM 308

79

Supplement to Series FM1 no. 34

Conception statistics

Conceptions for women resident in England and Wales, 2004

This publication is also available at the National Statistics
website: www.statistics.gov.uk

For any other use of this material please apply for a Click-Use
Licence for core material at www.opsi.gov.uk/click-use/system/
online/pLogin.asp or by writing to:
Office of Public Sector Information
Information Policy Team
St Clements House
2–16 Colegate
Norwich NR3 1BQ
Fax: 01603 723000
E-mail: hmsolicensing@cabinet-office.x.gsi.gov.uk

Contact points
For enquiries about this publication, contact
Vital Statistics Outputs Branch
Tel: 01329 813758
E-mail: vsob@ons.gsi.gov.uk

For enquiries, about abortions statistics, contact
Department of Health
Tel: 020 7972 5537
E-mail: abortion.statistics@dh.gsi.gov.uk

For general enquiries, contact the National Statistics
Customer Contact Centre on: 0845 601 3034
(minicom: 01633 812399)
E-mail: info@statistics.gsi.gov.uk
Fax: 01633 652747
Post: Room 1015, Government Buildings,
Cardiff Road, Newport NP10 8XG

You can also find National Statistics on the Internet at:
www.statistics.gov.uk

About the Office for National Statistics
The Office for National Statistics (ONS) is the government
agency responsible for compiling, analysing and disseminating
many of the United Kingdom's economic, social and
demographic statistics, including the retail prices index, trade
figures and labour market data, as well as the periodic census
of the population and health statistics. It is also the agency that
administers the statutory registration of births, marriages and
deaths in England and Wales. The Director of ONS is also the
National Statistician and the Registrar General for England and
Wales.

A National Statistics publication
National Statistics are produced to high professional standards
set out in the National Statistics Code of Practice. They undergo
regular quality assurance reviews to ensure that they meet
customer needs. They are produced free from any political
influence.

Contents

List of main tables and appendices

1 Introduction

Conception statistics 2004 presents final statistics on conceptions for women usually resident in England and Wales. This is a supplement to series FM1 no. 34.[1] Previously, conception statistics were published in the annual volume *Birth statistics* (series FM1). Provisional figures for 2004 were presented in *Health Statistics Quarterly 29.*[2]

This supplement is produced by the Office for National Statistics (ONS). It is published under the National Statistics logo, the designation guaranteeing that those outputs have been produced to high professional standards set out in a Code of Practice, and have been produced free from political influence.

1.1 Tables in this supplement

The tables presented in this supplement are numbered Tables 12.1-12.10. This numbering convention has been adopted because these tables are a continuation of the tables published in *Birth statistics 2005.*[1]

For brevity, the time series shown here have been limited to a run of 11 years at most. Figures for earlier years are shown in earlier volumes of *Birth statistics.*

1.2 Data analysed in this supplement

Conception statistics bring together records of birth registrations collected under the Births and Deaths Registration Act 1953 and of abortions under the Abortions Act 1967.

They include all the pregnancies of women usually resident in England and Wales which lead to one of the following outcomes:
(i) a maternity at which one or more live births or stillbirths occur, which is registered in England and Wales;
(ii) a termination of a pregnancy by abortion under the 1967 Act, in England and Wales.

Pregnancies which lead to spontaneous abortions (i.e. miscarriages) are not included. Maternities which result in one or more live births or stillbirths are counted once only.

The registration of life events (births, deaths and marriages) is a service carried out by the Local Registration Service in partnership with the General Register Office (GRO) in Southport, which is part of ONS. Most of the information, for both live births and stillbirths, is supplied to registrars by one or other, or both of the parents. For stillbirths, details of cause of death, duration of pregnancy and weight of foetus are supplied on a certificate or notification by the doctor or midwife either present at the birth, or who examined the body. The certificate or notification is then taken by the informant to a registrar.

Further details on the information collected at birth or stillbirth can be found in section 1.2 of *Birth statistics 2005.*[1]

Information on abortions is derived from notifications supplied under the Abortions Act 1967. These are sent by registered practitioners to the Chief Medical Officer of the Department of Health, or to the Chief Medical Officer of the National Assembly for Wales. The details supplied include the woman's date of birth, marital status, and usual residence. Further details may be found in the ONS volume of *Abortions statistics*[3] (produced until 2001) and the Department of Health Statistical Bulletin *Abortion statistics, England and Wales: 2005.*[4]

1.3 Issues affecting the quality of the data in this supplement

Abortions data

Under arrangements made following implementation of the Abortion Act 1967, the Office for National Statistics and its predecessors processed and analysed the abortion notification forms (HSA4) sent to the Chief Medical Officers of England and Wales. From 1st April 2002, responsibility was transferred to the Department of Health where a new system was introduced to process the new abortion notification forms that were made available from 18th April 2002.

The collection of marital status information was affected in the changeover between the old and the new abortion notification form. There were some discontinuities in the recording and coding of marital status in the 2002 abortions data following the introduction of the new form. For further details see section 1.3 of *Conception statistics 2002.*[5]

Maternities data

Maternities data included in this supplement have been derived from the ONS Births database. On a small percentage of maternities records (1 per cent of 2004 conceptions) mother's date of birth was not present, and in these cases mother's age at conception was calculated from an imputed mother's age at maternity. For further information on imputation and other issues affecting the quality of birth statistics see section 1.3 of *Birth Statistics 2005.*[1]

1.4 Associated publications and the National Statistics website

The National Statistics website (www.statistics.gov.uk) provides a comprehensive source of freely available vital statistics and ONS products. More information on the National Statistics website can be obtained from ONS (section 2.8). Detailed conception figures, including quarterly conceptions to women aged under 18, are published regularly on the National Statistics website.

Figures for UK countries

In Scotland, figures for conceptions are available for women aged under 20. The definition of conception in Scotland includes maternities (pregnancies ending in a live birth or stillbirth) and pregnancies resulting in a therapeutic abortion or miscarriage which required hospital inpatient or day-case treatment. The figures are available on the ISD website www.isdscotland.org. There are no comparable conception figures for Northern Ireland.

Population Trends and Health Statistics Quarterly

Up to 1998 ONS published annual data in Monitors, known as the FM1 series for conceptions and live births. These contained basic information on annual conceptions and live birth registrations, and were issued soon after the data became available. However, these publications have been discontinued and since 1999 these data have appeared in Reports issued in the quarterly journal *Population Trends*. Since the beginning of 1999, ONS has published two quarterly journals: *Population Trends,* which now has an emphasis on population and demography, covering most fertility topics, and *Health Statistics Quarterly,* covering mortality and health topics, and some other fertility data. The annual Report on conceptions by local and health authority areas is published in *Health Statistics Quarterly.* Quarterly reference tables on conceptions are published in both *Population Trends* and *Health Statistics Quarterly.*

1.5 Other publications

Some other recent background publications on conceptions are listed below. Most are from the journals *Population Trends* and *Health Statistics Quarterly,* and where not available on the National Statistics website may be obtained from ONS (section 2.8).

Conceptions
- Griffiths, C and Kirby, L. (2000). 'Geographic variations in conceptions to women aged under 18 in Great Britain during the 1990s'. *Population Trends* 102, pp 13-23.
- Botting, B and Dunnell, K. (2000). 'Trends in fertility and contraception in the last quarter of the 20th century'. *Population Trends* 100, pp 32-40.

- Wood, R, Botting, B and Dunnell, K. (1997). 'Trends in conceptions before and after the 1995 pill scare'. *Population Trends* 89, pp 5-12.
- Wood, R. (1996). 'Subnational variations in conceptions'. *Population Trends* 84, pp 21-27.
- Babb, P. (1993). 'Teenage conceptions and fertility in England and Wales, 1971-91'. *Population Trends* 74, pp 12-17.

Abortions
- Lancucki, L and Ruddock, V. (2001). 'The calculation of abortion rates for England and Wales'. *Health Statistics Quarterly* 10, pp 25-32.

2 Notes and definitions

2.1 Base populations

The population figures (shown in Appendices 1 and 2) which are used to calculate conception rates in this supplement, are mid-year estimates of the resident population of England and Wales based on the 2001 Census of Population. These estimates include members of HM and non-UK armed forces stationed in England and Wales, but exclude those stationed outside. ONS mid-year population estimates are updated figures using the most recent Census, allowing for births, deaths, net migration and ageing of the population.

In this supplement, the population estimates used for the calculation of fertility rates are the latest consistent estimates available at the time of its production and were published as follows :-

Population estimates by age and sex
- mid-2004 estimates - published on 20 December 2005.

Population estimates by marital status
- mid-2004 marital status estimates - published on 20 December 2005.

Further details about population estimates can be found at the National Statistics website (www.statistics.gov.uk/popest).

2.2 Areal coverage

Conception figures include only live births, stillbirths and abortions that occurred in England and Wales to women usually resident in England and Wales. Numbers and rates of conceptions are given by mother's usual area of residence based on Local Authority boundaries as at 31 December 2004 and Health Authority boundaries as at 1 April 2005. The postcode of the woman's address at the time of the maternity or abortion was used to assign the local authority and health authority of residence at the time of the conception. Direct comparisons with conceptions data by area published in previous years are not always possible because of boundary changes.

2.3 Estimating the date of conception

Information on the exact date of conception cannot be obtained from the registration details for either births or abortions. The date is estimated as follows:

Maternities (one or more live births)
The date of conception is estimated as 38 weeks before the date of birth. The average duration of pregnancy, or gestation period, measured between the first day of the last menstrual period and the date of birth, is 40 weeks; conception occurs on average 14 days after the first day of the last menstrual period.

Maternities (all stillbirths)
The date of conception is calculated as the date of birth less the stated gestation period.

Abortions under the 1967 Act
For conceptions in 1980 and earlier years, the date of conception is taken as the date of the start of the last menstrual period, plus 14 days. For conceptions in 1981 and subsequent years, it is taken as the date of termination less the stated gestation period, plus 14 days.

2.4 Estimating a woman's age at conception

A woman's age at conception is estimated from her date of birth, as stated on the birth registration or abortion notification, together with the estimated date of conception. In the small number of cases of maternities for which the woman's date of birth was not stated, an age is imputed using the date of birth stated on a previous comparable record. In the small number of cases of abortions where the woman's date of birth was not recorded, the case is not used for statistical analysis.

The woman's age at conception is calculated as the interval in complete years between her date of birth and the date she conceived (estimated as described in section 2.3 above). However, where conception occurs in the same calendar month as the woman's birthday, this can result in the estimated age of mother at conception being either a year too low – if the baby is born live after less than 38 weeks – or a year too high – if the baby is born live at over 38 weeks. The method for estimating a woman's age at conception in such cases was revised in 1999 to take into account the mother's *day* of birth and *day* of conception in addition to the month and year as used previously. This resulted in a revision to figures published before 1999. All figures in this supplement are based on this revised methodology. A full explanation can be found in *Birth statistics 1998*.[6]

2.5 Conceptions within and outside marriage

A birth within marriage is that of a child born to parents who were lawfully married to each other either:
(a) at the date of the child's birth, or
(b) when the child was conceived, even if they later divorced or the father died before the child's birth.

Only for a birth within marriage will the registrar enter on the draft entry in the register confidential particulars relating to the date of the parents' marriage, whether the mother has been married more than once, and the number of the mother's previous liveborn and stillborn children . Births occurring outside marriage may be registered either jointly or solely. A joint registration records details of both parents, and requires them both to be present. A sole registration records only the mother's details.

Conceptions outside marriage which lead to maternities within marriage are restricted here to those where birth occurred within 8 months (35 weeks) of marriage. However, the date of conception and age of woman at conception are estimated from a date 38 weeks before birth, as in the case of all other conceptions leading to live births.

For abortions coded on the old form HSA4 originating from ONS, the categories of marital status were (a) single, (b) married, (c) widowed, (d) divorced and (e) separated. For abortions coded on the new form HSA4 originating from Department of Health, marital status was reclassified as (a) single no partner, (b) married, (c) single with partner, (d) single not known, (e) divorced, (f) widowed and (g) separated. To calculate rates it is necessary to include separated with married because there are no population estimates for the 'separated' marital status category.

2.6 Legislation

The main statutes concerning birth registration and provision of information on births are given below:

- **Census Act 1920,** which in Section 5 provides for the collection and publication of statistical information on the population by the Registrar General.
- **Population Statistics Act 1938,** which deals with the statistical information collected at registration.
- **Births and Deaths Registration Act 1953,** which covers all aspects of the registration of births and deaths.
- **Registration Service Act 1953,** which in Section 19 requires the Registrar General to provide annual abstracts of live births and stillbirths.
- **Population Statistics Act 1960,** which makes further provision for collecting statistical detail at registration.
- **Abortions Act 1967,** which permits termination of pregnancy by a registered practitioner, subject to certain conditions.

- **Registration of Births, Deaths and Marriages Regulations 1968,** which added questions on father's and mother's place of birth to the details requested at registration.
- **National Health Service Act 1977,** which requires notification of a birth to the health authority where the birth occurred.
- **Human Fertilisation and Embryology Act 1990,** Section 37 made changes to the Abortion Act 1967.
- **Stillbirth (Definition) Act 1992,** which altered the definition of a stillbirth to 24 or more weeks completed gestation, instead of the previous definition of 28 or more weeks completed gestation.
- **Health Act 1999,** a section of which includes specific provision for the supply of information on individual births to the National Health Service.

2.7 Symbols and conventions

In this volume:
: denotes not appropriate
0 denotes less than 0.05
- denotes nil.
* denotes not available (to protect confidentiality)

Where data are not yet available, cells in tables are left blank. Rates calculated from fewer than 20 events are distinguished by italic type as a warning to users that their reliability as a measure may be affected by the small number of events. Data displaying conception statistics as counts and rates cannot be used to disclose information on abortions. Therefore to protect confidentiality for conceptions data, all counts lower than 5, and all rates based on fewer than 5 events have been suppressed. For conceptions leading to abortions, counts less than 10 and rates based on fewer than 10 events have been suppressed. For further details on protecting health statistics containing abortions data see the Disclosure Review for Health Statistics.[7]

Occasionally it has been necessary to apply a secondary suppression to avoid the possibility of disclosure by differencing. Figures in some tables in this publication may not add precisely due to rounding.

2.8 Further information

Requests for conceptions data, as well as background information on this volume, VS tables, and on data quality, should be made to:

Vital Statistics Outputs Branch
Office for National Statistics
Segensworth Road
Titchfield
Fareham
Hants PO15 5RR
Telephone: 01329 813758; Fax: 01329 813548
email: vsob@ons.gsi.gov.uk

References

1. ONS (2006) Birth statistics 2005, series FM1 no. 34.
2. ONS Report: Conceptions in England and Wales, 2004. *Health Statistics Quarterly* 29, pp 54-58.
3. ONS (2001) *Abortion statistics*, series AB no. 28.
4. Department of Health Statistical Bulletin Abortion Statistics, England and Wales: 2005.
    See **http://www.dh.gov.uk/Publications**
5. ONS (2005) Conception statistics 2002, supplement to series FM1 no. 32.
6. ONS (1998) Birth statistics 1998, series FM1 no. 27.
7. ONS (2005) Disclosure Review for Health Statistics, 1st Report - Guidance for Abortion Statistics.
    See **http://www.statistics.gov.uk/statbase/ Product.asp?vlnk =11988**

Table 12.1 Conceptions (numbers and percentages and rates): occurrence within/outside marriage and outcome, and age at conception, 1994-2004

England and Wales
Residents

Age of woman at conception and year of conception	Number of conceptions (thousands)			Percentage of all conceptions				Conception rates per 1,000 women in age-group[6]		
				Outcome		Occurrence				
	Total conceptions	Conceptions leading to maternities	Conceptions terminated by abortion	Leading to maternities	Terminated by abortion	Within marriage	Outside marriage	Total conceptions	Conceptions leading to maternities	Conceptions terminated by abortion
All ages[1]										
1994	**801.6**	645.5	156.0	80.5	19.5	54.4	45.6	**74.8**	60.2	14.6
1995	**790.3**	634.4	155.8	80.3	19.7	52.9	47.1	**73.8**	59.2	14.5
1996	**816.9**	647.2	169.7	79.2	20.8	51.0	49.0	**76.2**	60.4	15.8
1997	**800.4**	629.9	170.5	78.7	21.3	50.2	49.8	**74.6**	58.7	15.9
1998	**797.0**	619.6	177.4	77.7	22.3	48.8	51.2	**74.2**	57.7	16.5
1999	**774.0**	599.4	174.6	77.4	22.6	48.3	51.7	**71.9**	55.7	16.2
2000	**767.0**	592.6	174.4	77.3	22.7	47.7	52.3	**70.9**	54.8	16.1
2001	**763.7**	586.6	177.1	76.8	23.2	47.0	53.0	**70.3**	54.0	16.3
2002	**787.0**	609.9	177.2	77.5	22.5	46.5	53.5	**72.2**	55.9	16.2
2003	**806.8**	625.4	181.4	77.5	22.5	45.5	54.5	**73.7**	57.1	16.6
2004	**826.8**	641.7	185.1	77.6	22.4	45.1	54.9	**75.3**	58.4	16.8
Under 16[2]										
1994	**7.8**	3.9	3.9	49.7	50.3	0.4	99.6	**8.3**	4.1	4.2
1995	**8.1**	4.2	3.8	52.4	47.6	0.3	99.7	**8.6**	4.5	4.1
1996	**8.9**	4.5	4.4	50.8	49.2	0.2	99.8	**9.5**	4.8	4.7
1997	**8.3**	4.2	4.1	50.3	49.7	0.3	99.7	**8.9**	4.5	4.4
1998	**8.5**	4.0	4.4	47.6	52.4	0.3	99.7	**9.0**	4.3	4.7
1999	**7.9**	3.8	4.2	47.4	52.6	0.5	99.5	**8.3**	3.9	4.4
2000	**8.1**	3.7	4.4	46.0	54.0	0.3	99.7	**8.3**	3.8	4.5
2001	**7.9**	3.5	4.4	44.2	55.8	0.4	99.6	**8.0**	3.5	4.5
2002	**7.9**	3.5	4.4	44.4	55.6	0.4	99.6	**7.9**	3.5	4.4
2003	**8.0**	3.4	4.6	42.6	57.4	0.4	99.6	**8.0**	3.4	4.6
2004	**7.6**	3.3	4.4	42.8	57.2	0.5	99.5	**7.5**	3.2	4.3
Under 18[3]										
1994	**36.1**	21.7	14.3	60.2	39.8	3.0	97.0	**41.9**	25.3	16.7
1995	**37.9**	23.2	14.7	61.3	38.7	2.6	97.4	**41.9**	25.7	16.2
1996	**43.5**	26.1	17.4	60.0	40.0	2.3	97.7	**46.3**	27.8	18.6
1997	**43.4**	25.8	17.6	59.4	40.6	2.4	97.6	**45.9**	27.3	18.6
1998	**44.1**	25.6	18.5	58.0	42.0	2.3	97.7	**47.1**	27.4	19.8
1999	**42.0**	23.9	18.1	57.0	43.0	2.2	97.8	**45.1**	25.7	19.4
2000	**41.3**	23.1	18.3	55.8	44.2	2.2	97.8	**43.9**	24.5	19.4
2001	**41.0**	22.3	18.7	54.3	45.7	2.2	97.8	**42.7**	23.2	19.5
2002	**42.0**	22.9	19.0	54.7	45.3	2.2	97.8	**42.8**	23.4	19.4
2003	**42.2**	22.9	19.3	54.3	45.7	2.0	98.0	**42.3**	23.0	19.3
2004	**42.2**	23.0	19.2	54.4	45.6	1.8	98.2	**41.7**	22.7	19.0
Under 20[4]										
1994	**85.4**	55.7	29.6	65.3	34.7	8.2	91.8	**58.7**	38.3	20.4
1995	**86.6**	56.6	29.9	65.4	34.6	7.6	92.4	**58.7**	38.4	20.3
1996	**94.9**	60.5	34.4	63.8	36.2	6.8	93.2	**63.2**	40.3	22.9
1997	**96.0**	60.7	35.3	63.2	36.8	6.7	93.3	**62.6**	39.6	23.0
1998	**101.6**	63.2	38.4	62.2	37.8	6.6	93.4	**65.1**	40.5	24.6
1999	**98.8**	60.7	38.1	61.4	38.6	6.5	93.5	**63.1**	38.7	24.3
2000	**97.7**	59.2	38.4	60.7	39.3	6.6	93.4	**62.5**	37.9	24.6
2001	**96.0**	57.2	38.7	59.6	40.4	6.5	93.5	**60.8**	36.3	24.5
2002	**97.1**	58.3	38.8	60.1	39.9	6.3	93.7	**60.3**	36.3	24.1
2003	**98.6**	58.9	39.7	59.8	40.2	5.7	94.3	**59.8**	35.7	24.1
2004	**101.3**	60.7	40.6	59.9	40.1	5.4	94.6	**60.3**	36.1	24.2

1 Rates per 1,000 women aged 15-44.
2 Rates per 1,000 women aged 13-15.
3 Rates per 1,000 women aged 15-17.
4 Rates per 1,000 women aged 15-19.
6 See section 2.1.

Table 12.1 - *continued*

Age of woman at conception and year of conception	Number of conceptions (thousands)			Percentage of all conceptions				Conception rates per 1,000 women in age-group[6]		
	Total conceptions	Conceptions leading to maternities	Conceptions terminated by abortion	Outcome		Occurrence		Total conceptions	Conceptions leading to maternities	Conceptions terminated by abortion
				Leading to maternities	Terminated by abortion	Within marriage	Outside marriage			
20–24										
1994	**190.4**	145.8	44.6	76.6	23.4	37.7	62.3	**107.3**	82.2	25.1
1995	**181.1**	137.4	43.8	75.8	24.2	35.3	64.7	**105.8**	80.2	25.6
1996	**179.8**	133.5	46.3	74.3	25.7	32.5	67.5	**110.1**	81.8	28.4
1997	**167.3**	122.6	44.6	73.3	26.7	31.3	68.7	**107.2**	78.6	28.6
1998	**163.3**	117.8	45.4	72.2	27.8	29.9	70.1	**107.7**	77.7	30.0
1999	**157.6**	112.6	45.0	71.5	28.5	28.9	71.1	**103.9**	74.3	29.6
2000	**159.0**	112.5	46.4	70.8	29.2	28.3	71.7	**103.2**	73.1	30.1
2001	**161.6**	113.6	48.1	70.3	29.7	27.9	72.1	**102.5**	72.0	30.5
2002	**167.8**	119.5	48.3	71.2	28.8	27.1	72.9	**104.6**	74.5	30.1
2003	**175.3**	124.5	50.7	71.0	29.0	26.0	74.0	**107.1**	76.1	31.0
2004	**181.3**	128.9	52.4	71.1	28.9	25.5	74.5	**108.9**	77.4	31.5
25–29										
1994	**261.8**	224.3	37.5	85.7	14.3	64.6	35.4	**128.3**	109.9	18.4
1995	**250.3**	213.2	37.1	85.2	14.8	62.6	37.4	**124.8**	106.3	18.5
1996	**252.6**	213.1	39.5	84.4	15.6	60.4	39.6	**127.6**	107.6	20.0
1997	**242.6**	202.8	39.8	83.6	16.4	58.9	41.1	**124.7**	104.2	20.5
1998	**232.4**	192.6	39.8	82.9	17.1	57.3	42.7	**122.2**	101.3	20.9
1999	**218.5**	180.2	38.3	82.5	17.5	56.5	43.5	**118.0**	97.4	20.7
2000	**209.3**	172.3	37.0	82.3	17.7	55.9	44.1	**115.7**	95.2	20.4
2001	**199.3**	162.7	36.6	81.6	18.4	54.7	45.3	**114.2**	93.3	21.0
2002	**199.4**	163.8	35.6	82.1	17.9	54.2	45.8	**119.1**	97.8	21.3
2003	**199.8**	164.1	35.8	82.1	17.9	53.1	46.9	**122.8**	100.8	22.0
2004	**205.1**	167.8	37.3	81.8	18.2	52.5	47.5	**126.2**	103.2	22.9
30–34										
1994	**185.0**	159.9	25.1	86.4	13.6	72.1	27.9	**92.1**	79.6	12.5
1995	**190.3**	164.5	25.8	86.4	13.6	70.9	29.1	**92.8**	80.2	12.6
1996	**200.0**	171.8	28.2	85.9	14.1	69.6	30.4	**96.3**	82.8	13.6
1997	**200.9**	172.3	28.6	85.8	14.2	68.9	31.1	**96.2**	82.5	13.7
1998	**201.4**	171.4	30.0	85.1	14.9	67.7	32.3	**96.8**	82.4	14.4
1999	**197.1**	168.0	29.1	85.3	14.7	67.2	32.8	**95.3**	81.2	14.1
2000	**195.3**	167.0	28.3	85.5	14.5	66.8	33.2	**95.3**	81.5	13.8
2001	**196.7**	168.0	28.7	85.4	14.6	66.2	33.8	**96.7**	82.6	14.1
2002	**204.3**	175.9	28.4	86.1	13.9	65.4	34.6	**101.6**	87.5	14.1
2003	**209.0**	180.6	28.4	86.4	13.6	64.7	35.3	**105.9**	91.5	14.4
2004	**209.6**	181.9	27.7	86.8	13.2	64.9	35.1	**109.4**	94.9	14.5
35–39										
1994	**66.2**	52.2	14.0	78.9	21.1	69.5	30.5	**37.6**	29.7	7.9
1995	**68.7**	54.5	14.2	79.3	20.7	68.2	31.8	**38.1**	30.2	7.9
1996	**75.5**	59.5	16.0	78.8	21.2	67.2	32.8	**40.7**	32.1	8.6
1997	**78.9**	62.3	16.6	79.0	21.0	66.3	33.7	**41.4**	32.7	8.7
1998	**82.9**	65.0	17.8	78.5	21.5	65.3	34.7	**42.4**	33.3	9.1
1999	**86.0**	67.7	18.2	78.8	21.2	64.9	35.1	**42.9**	33.8	9.1
2000	**88.7**	70.5	18.2	79.5	20.5	64.3	35.7	**43.2**	34.3	8.8
2001	**92.2**	73.4	18.9	79.6	20.4	63.3	36.7	**44.3**	35.2	9.1
2002	**98.9**	79.6	19.3	80.5	19.5	62.1	37.9	**47.0**	37.9	9.2
2003	**103.1**	83.7	19.4	81.1	18.9	61.0	39.0	**49.1**	39.8	9.3
2004	**106.8**	87.2	19.6	81.7	18.3	60.9	39.1	**51.0**	41.7	9.4

6 See section 2.1.

Table 12.1 - *continued*

Age of woman at conception and year of conception	Number of conceptions (thousands)			Percentage of all conceptions				Conception rates per 1,000 women in age-group[6]		
	Total con- ceptions	Conceptions leading to maternities	Conceptions terminated by abortion	Outcome		Occurrence		**Total con- ceptions**	Conceptions leading to maternities	Conceptions terminated by abortion
				Leading to maternities	Terminated by abortion	Within marriage	Outside marriage			
40 and over[5]										
1994	**12.9**	7.6	5.3	59.1	40.9	66.4	33.6	**7.6**	4.5	3.1
1995	**13.2**	8.2	5.0	62.0	38.0	64.4	35.6	**7.9**	4.9	3.0
1996	**14.1**	8.8	5.3	62.4	37.6	63.2	36.8	**8.4**	5.3	3.2
1997	**14.7**	9.1	5.6	62.0	38.0	62.8	37.2	**8.7**	5.4	3.3
1998	**15.4**	9.6	5.8	62.1	37.9	61.8	38.2	**8.9**	5.5	3.4
1999	**16.0**	10.1	5.9	63.0	37.0	60.5	39.5	**9.1**	5.7	3.4
2000	**17.0**	11.0	6.0	64.6	35.4	60.7	39.3	**9.4**	6.1	3.3
2001	**17.8**	11.7	6.2	65.4	34.6	58.6	41.4	**9.6**	6.3	3.3
2002	**19.6**	12.8	6.8	65.4	34.6	56.3	43.7	**10.3**	6.7	3.6
2003	**20.9**	13.7	7.3	65.3	34.7	54.7	45.3	**10.7**	7.0	3.7
2004	**22.8**	15.3	7.5	67.0	33.0	53.8	46.2	**11.4**	7.6	3.7

5   Rates per 1,000 women aged 40–44.
6   See section 2.1.

Table 12.2 Teenage conceptions (numbers and rates): outcome and age of woman (single years) at conception, 1994-2004

England and Wales
Residents

Age of woman at conception and year of conception	Number of conceptions			Conception rates per 1,000 women in age-group			Age of woman at conception and year of conception	Number of conceptions			Conception rates per 1,000 women in age-group		
	Total conceptions	Conceptions leading to maternities	Conceptions terminated by abortion	**Total conceptions**	Conceptions leading to maternities	Conceptions terminated by abortion		**Total conceptions**	Conceptions leading to maternities	Conceptions terminated by abortion	**Total conceptions**	Conceptions leading to maternities	Conceptions terminated by abortion
Under 16[1]							**16**						
1994	**7,795**	3,875	3,920	**8.3**	4.1	4.2	1994	**11,336**	6,833	4,503	**40.4**	24.4	16.0
1995	**8,051**	4,218	3,833	**8.6**	4.5	4.1	1995	**12,382**	7,668	4,714	**40.6**	25.1	15.5
1996	**8,857**	4,498	4,359	**9.5**	4.8	4.7	1996	**14,284**	8,606	5,678	**45.0**	27.1	17.9
1997	**8,271**	4,164	4,107	**8.9**	4.5	4.4	1997	**14,058**	8,364	5,694	**44.5**	26.5	18.0
1998	**8,452**	4,023	4,429	**9.0**	4.3	4.7	1998	**13,802**	8,078	5,724	**44.5**	26.1	18.5
1999	**7,945**	3,762	4,183	**8.3**	3.9	4.4	1999	**13,334**	7,489	5,845	**43.0**	24.2	18.9
2000	**8,114**	3,730	4,384	**8.3**	3.8	4.5	2000	**13,153**	7,299	5,854	**42.3**	23.5	18.8
2001	**7,903**	3,492	4,411	**8.0**	3.5	4.5	2001	**13,103**	7,106	5,997	**40.4**	21.9	18.5
2002	**7,875**	3,497	4,378	**7.9**	3.5	4.4	2002	**13,475**	7,283	6,192	**41.5**	22.4	19.1
2003	**8,024**	3,415	4,609	**8.0**	3.4	4.6	2003	**13,303**	7,210	6,093	**40.1**	21.7	18.4
2004	**7,615**	3,263	4,352	**7.5**	3.2	4.3	2004	**13,636**	7,415	6,221	**40.0**	21.7	18.2
Under 14[2]							**17**						
1994	**397**	167	230	**1.3**	0.5	0.7	1994	**16,960**	11,036	5,924	**61.4**	40.0	21.5
1995	**382**	150	232	**1.2**	0.5	0.8	1995	**17,447**	11,321	6,126	**62.0**	40.3	21.8
1996	**451**	192	259	**1.5**	0.6	0.8	1996	**20,349**	12,975	7,374	**66.6**	42.5	24.1
1997	**365**	149	216	**1.2**	0.5	0.7	1997	**21,029**	13,242	7,787	**66.1**	41.6	24.5
1998	**423**	170	253	**1.3**	0.5	0.8	1998	**21,865**	13,503	8,362	**69.1**	42.7	26.4
1999	**406**	174	232	**1.3**	0.5	0.7	1999	**20,749**	12,686	8,063	**66.8**	40.9	26.0
2000	**397**	161	236	**1.2**	0.5	0.7	2000	**20,081**	12,061	8,020	**65.0**	39.0	26.0
2001	**400**	179	221	**1.2**	0.5	0.7	2001	**19,984**	11,652	8,332	**63.8**	37.2	26.6
2002	**390**	149	241	**1.2**	0.4	0.7	2002	**20,601**	12,162	8,439	**63.1**	37.3	25.9
2003	**334**	128	206	**1.0**	0.4	0.6	2003	**20,835**	12,284	8,551	**63.6**	37.5	26.1
2004	**337**	128	209	**1.0**	0.4	0.6	2004	**20,947**	12,289	8,658	**62.6**	36.7	25.9
14							**18**						
1994	**1,938**	804	1,134	**6.1**	2.5	3.6	1994	**22,614**	15,389	7,225	**78.5**	53.4	25.1
1995	**1,834**	878	956	**5.8**	2.8	3.0	1995	**22,402**	15,073	7,329	**81.1**	54.6	26.5
1996	**1,961**	838	1,123	**6.4**	2.7	3.6	1996	**24,150**	15,912	8,238	**86.1**	56.7	29.4
1997	**1,964**	866	1,098	**6.4**	2.8	3.6	1997	**25,618**	16,828	8,790	**84.2**	55.3	28.9
1998	**1,988**	821	1,167	**6.4**	2.6	3.8	1998	**27,939**	17,969	9,970	**88.3**	56.8	31.5
1999	**1,866**	785	1,081	**5.8**	2.4	3.3	1999	**26,627**	16,990	9,637	**84.6**	54.0	30.6
2000	**1,890**	790	1,100	**5.9**	2.4	3.4	2000	**26,180**	16,529	9,651	**85.9**	54.2	31.6
2001	**1,890**	729	1,161	**5.8**	2.2	3.5	2001	**25,746**	16,069	9,677	**83.6**	52.2	31.4
2002	**1,858**	719	1,139	**5.5**	2.1	3.4	2002	**25,910**	16,314	9,596	**81.9**	51.6	30.3
2003	**1,888**	676	1,212	**5.7**	2.0	3.6	2003	**26,610**	16,569	10,041	**80.6**	50.2	30.4
2004	**1,754**	649	1,105	**5.2**	1.9	3.3	2004	**27,373**	17,031	10,342	**82.5**	51.3	31.2
15							**19**						
1994	**5,460**	2,904	2,556	**17.9**	9.5	8.4	1994	**26,647**	18,590	8,057	**87.3**	60.9	26.4
1995	**5,835**	3,190	2,645	**18.4**	10.1	8.3	1995	**26,305**	18,359	7,946	**89.5**	62.5	27.0
1996	**6,445**	3,468	2,977	**20.4**	11.0	9.4	1996	**27,233**	18,494	8,739	**96.5**	65.5	31.0
1997	**5,942**	3,149	2,793	**19.2**	10.2	9.0	1997	**27,031**	18,110	8,921	**94.5**	63.3	31.2
1998	**6,041**	3,032	3,009	**19.5**	9.8	9.7	1998	**29,569**	19,643	9,926	**95.9**	63.7	32.2
1999	**5,673**	2,803	2,870	**18.2**	9.0	9.2	1999	**30,132**	19,734	10,398	**94.1**	61.6	32.5
2000	**5,827**	2,779	3,048	**18.1**	8.6	9.5	2000	**30,136**	19,619	10,517	**95.5**	62.1	33.3
2001	**5,613**	2,584	3,029	**17.4**	8.0	9.4	2001	**29,234**	18,913	10,321	**94.5**	61.1	33.4
2002	**5,627**	2,629	2,998	**17.1**	8.0	9.1	2002	**29,246**	19,084	10,162	**93.7**	61.1	32.5
2003	**5,802**	2,611	3,191	**17.2**	7.7	9.4	2003	**29,820**	19,433	10,387	**92.7**	60.4	32.3
2004	**5,524**	2,486	3,038	**16.5**	7.4	9.1	2004	**31,691**	20,684	11,007	**94.3**	61.6	32.8

1 Rates per 1,000 women aged 13-15.
2 Rates per 1,000 women aged 13.

Table 12.3 Conceptions within marriage (numbers and percentages): outcome and marriage order, and age of woman at conception, 1994-2004

England and Wales
Residents

Age of woman at conception and year of conception	Total number of conceptions within marriage (000s)	Percentage of conceptions within marriage				Age of woman at conception and year of conception	Total number of conceptions within marriage (000s)	Percentage of conceptions within marriage			
		Leading to maternities			Terminated by abortion			Leading to maternities			Terminated by abortion
		All marriages	First marriage	Second or later marriage				All marriages	First marriage	Second or later marriage	
All ages						**25–29**					
1994	435.9	92.1	85.8	6.4	7.9	1994	169.2	94.6	90.5	4.1	5.4
1995	417.7	92.1	85.7	6.4	7.9	1995	156.8	94.5	90.6	3.9	5.5
1996	416.3	91.7	85.4	6.4	8.3	1996	152.6	94.3	90.5	3.8	5.7
1997	401.5	91.5	85.2	6.3	8.5	1997	142.9	94.1	90.6	3.5	5.9
1998	388.8	91.2	85.1	6.1	8.8	1998	133.2	94.0	90.7	3.3	6.0
1999	373.5	91.2	85.4	5.9	8.8	1999	123.5	94.0	90.9	3.1	6.0
2000	366.2	91.5	85.8	5.6	8.5	2000	116.9	93.9	91.1	2.9	6.1
2001	359.2	91.4	86.0	5.4	8.6	2001	108.9	93.7	91.2	2.5	6.3
2002	365.7	92.2	86.9	5.2	7.8	2002	108.1	94.2	91.8	2.3	5.8
2003	367.0	92.8	87.8	5.0	7.2	2003	106.1	94.5	92.4	2.1	5.5
2004	372.7	93.0	88.3	4.7	7.0	2004	107.8	94.4	92.5	2.0	5.6
Under 18						**30–34**					
1994	1.1	93.9	93.9	0.1	6.1	1994	133.5	92.1	83.3	8.8	7.9
1995	1.0	93.7	93.7	0.0	6.3	1995	134.9	92.3	83.6	8.8	7.7
1996	1.0	94.1	94.1	0.0	5.9	1996	139.2	92.2	83.9	8.2	7.8
1997	1.0	93.0	93.0	0.0	7.0	1997	138.4	92.3	84.3	7.9	7.7
1998	1.0	91.1	91.1	0.0	8.9	1998	136.3	91.9	84.5	7.4	8.1
1999	0.9	92.0	91.9	0.1	8.0	1999	132.4	92.3	85.2	7.1	7.7
2000	0.9	92.3	92.3	0.0	7.7	2000	130.4	92.8	86.1	6.6	7.2
2001	0.9	94.3	94.1	0.1	5.7	2001	130.1	92.8	86.5	6.4	7.2
2002	0.9	91.2	91.1	0.1	8.8	2002	133.6	93.7	87.8	5.9	6.3
2003	0.8	92.2	92.2	0.0	7.8	2003	135.3	94.4	89.0	5.4	5.6
2004	0.8	90.6	90.5	0.1	9.4	2004	136.0	94.6	89.7	4.9	5.4
Under 20						**35–39**					
1994	7.0	94.8	94.6	0.3	5.2	1994	46.0	84.1	69.2	14.9	15.9
1995	6.6	94.8	94.6	0.2	5.2	1995	46.9	84.5	70.0	14.5	15.5
1996	6.5	94.3	94.2	0.1	5.7	1996	50.8	84.0	69.9	14.1	16.0
1997	6.4	94.0	93.8	0.1	6.0	1997	52.3	84.6	70.8	13.8	15.4
1998	6.7	93.2	93.1	0.1	6.8	1998	54.1	84.7	71.5	13.1	15.3
1999	6.4	93.0	92.8	0.1	7.0	1999	55.8	85.2	72.9	12.3	14.8
2000	6.5	93.2	93.1	0.1	6.8	2000	57.0	86.2	74.4	11.8	13.8
2001	6.3	94.0	93.8	0.2	6.0	2001	58.4	86.3	75.0	11.3	13.7
2002	6.1	93.0	92.9	0.1	7.0	2002	61.4	88.1	76.9	11.1	11.9
2003	5.6	93.3	93.2	0.1	6.7	2003	62.9	89.4	78.6	10.8	10.6
2004	5.5	93.0	92.9	0.1	7.0	2004	65.0	90.1	79.9	10.2	9.9
20–24						**40 and over**					
1994	71.8	94.6	93.4	1.2	5.4	1994	8.5	63.1	47.9	15.2	36.9
1995	64.0	94.4	93.3	1.1	5.6	1995	8.5	66.3	50.1	16.2	33.7
1996	58.4	94.0	92.8	1.2	6.0	1996	8.9	66.8	51.1	15.7	33.2
1997	52.3	93.8	92.7	1.1	6.2	1997	9.2	66.2	50.1	16.1	33.8
1998	48.9	93.4	92.5	0.9	6.6	1998	9.5	66.2	50.3	15.9	33.8
1999	45.6	93.1	92.2	0.9	6.9	1999	9.7	67.5	52.7	14.8	32.5
2000	45.0	92.8	92.0	0.8	7.2	2000	10.3	69.7	54.4	15.3	30.3
2001	45.1	92.7	91.9	0.8	7.3	2001	10.5	71.0	56.8	14.2	29.0
2002	45.4	92.9	92.2	0.7	7.1	2002	11.0	72.8	58.4	14.4	27.2
2003	45.7	93.2	92.5	0.6	6.8	2003	11.5	74.2	59.6	14.6	25.8
2004	46.2	93.4	92.7	0.7	6.6	2004	12.2	76.4	62.3	14.1	23.6

Table 12.4 Conceptions outside marriage (numbers and percentages): outcome and whether sole or joint registration, and age of woman at conception, 1994-2004

England and Wales
Residents

Age of woman at conception and year of conception	Total number of conceptions outside marriage (000s)	Percentage of conceptions outside marriage					
		Leading to maternities					Terminated by abortion
		Outside marriage			Within marriage		
		Total	Sole[1]	Joint[2]			
All ages							
1994	**365.6**	59.5	13.1	46.4	7.2		33.3
1995	**372.5**	60.2	13.3	46.9	6.9		32.9
1996	**400.6**	59.8	12.9	46.9	6.4		33.8
1997	**398.9**	59.5	12.4	47.1	6.3		34.2
1998	**408.2**	59.1	11.9	47.1	5.9		35.1
1999	**400.4**	59.1	11.5	47.6	5.5		35.4
2000	**400.8**	59.1	11.0	48.1	5.2		35.7
2001	**404.4**	58.9	10.6	48.3	5.0		36.1
2002	**421.3**	59.8	10.5	49.3	4.9		35.2
2003	**439.8**	59.9	10.2	49.7	4.9		35.2
2004	**454.1**	60.4	10.0	50.5	4.6		35.0
Under 16							
1994	**7.8**	49.2	23.5	25.7	0.4		50.5
1995	**8.0**	51.6	23.5	28.1	0.7		47.8
1996	**8.8**	50.1	22.8	27.2	0.6		49.3
1997	**8.2**	49.8	22.8	27.0	0.4		49.8
1998	**8.4**	47.1	20.9	26.1	0.4		52.5
1999	**7.9**	46.8	21.2	25.6	0.4		52.9
2000	**8.1**	45.4	19.6	25.7	0.5		54.2
2001	**7.9**	43.6	19.3	24.3	0.4		56.0
2002	**7.8**	43.9	19.1	24.8	0.3		55.8
2003	**8.0**	42.1	17.9	24.2	0.2		57.7
2004	**7.6**	42.4	17.4	25.0	0.2		57.4
Under 18							
1994	**35.0**	57.1	20.6	36.6	2.1		40.8
1995	**36.9**	58.5	21.2	37.4	1.9		39.6
1996	**42.5**	57.5	20.3	37.1	1.7		40.8
1997	**42.3**	57.0	19.9	37.1	1.6		41.4
1998	**43.1**	55.7	19.3	36.5	1.5		42.7
1999	**41.1**	54.8	18.7	36.2	1.3		43.8
2000	**40.4**	53.8	17.4	36.4	1.2		45.0
2001	**40.1**	52.1	16.7	35.4	1.2		46.6
2002	**41.0**	52.8	16.7	36.0	1.1		46.1
2003	**41.3**	52.6	16.3	36.4	0.9		46.4
2004	**41.4**	52.9	15.6	37.3	0.9		46.2
Under 20							
1994	**78.3**	59.0	18.6	40.5	3.6		37.4
1995	**80.0**	59.8	19.0	40.8	3.2		37.0
1996	**88.4**	58.5	18.3	40.2	3.0		38.5
1997	**89.6**	58.4	18.0	40.3	2.7		39.0
1998	**94.9**	57.4	17.2	40.2	2.6		40.0
1999	**92.4**	56.8	16.4	40.4	2.4		40.8
2000	**91.2**	56.2	16.0	40.2	2.2		41.7
2001	**89.7**	55.1	15.1	40.0	2.1		42.8
2002	**91.0**	55.9	15.1	40.8	2.0		42.1
2003	**93.0**	55.9	14.6	41.2	1.9		42.3
2004	**95.8**	56.3	14.1	42.2	1.7		42.0

Age of woman at conception and year of conception	Total number of conceptions outside marriage (000s)	Percentage of conceptions outside marriage					
		Leading to maternities					Terminated by abortion
		Outside marriage			Within marriage		
		Total	Sole[1]	Joint[2]			
20–24							
1994	**118.6**	58.6	12.7	45.9	7.1		34.3
1995	**117.1**	59.0	12.9	46.1	6.7		34.3
1996	**121.4**	58.7	12.6	46.0	6.1		35.2
1997	**115.0**	58.1	12.2	45.9	5.9		36.0
1998	**114.4**	57.6	11.8	45.8	5.5		36.9
1999	**112.0**	57.6	11.5	46.1	5.0		37.3
2000	**113.9**	57.2	10.9	46.3	4.8		37.9
2001	**116.6**	57.2	10.6	46.6	4.4		38.4
2002	**122.3**	58.8	10.8	48.0	4.4		36.8
2003	**129.6**	58.9	10.3	48.6	4.3		36.7
2004	**135.1**	59.5	10.1	49.4	3.9		36.5
25–29							
1994	**92.6**	60.6	11.1	49.6	8.7		30.7
1995	**93.5**	61.1	11.1	50.1	8.5		30.4
1996	**100.0**	61.2	10.7	50.5	7.9		30.9
1997	**99.6**	60.6	10.2	50.4	8.0		31.5
1998	**99.2**	60.3	9.6	50.8	7.6		32.1
1999	**94.9**	60.4	9.3	51.1	7.1		32.4
2000	**92.4**	60.8	9.0	51.8	6.8		32.4
2001	**90.4**	60.6	8.7	51.9	6.5		32.9
2002	**91.3**	61.4	8.8	52.6	6.4		32.2
2003	**93.7**	61.6	8.7	52.9	6.5		31.9
2004	**97.4**	61.8	8.4	53.4	6.1		32.1
30–34							
1994	**51.5**	62.2	11.0	51.2	9.5		28.3
1995	**55.4**	63.1	11.0	52.1	9.0		27.9
1996	**60.8**	62.8	10.7	52.2	8.7		28.5
1997	**62.6**	62.7	9.7	53.0	8.7		28.5
1998	**65.1**	62.8	9.4	53.5	7.9		29.2
1999	**64.7**	63.1	8.8	54.3	7.7		29.2
2000	**64.9**	63.7	8.5	55.2	7.3		29.1
2001	**66.5**	64.1	8.2	55.9	6.9		29.0
2002	**70.6**	64.7	8.0	56.7	7.1		28.3
2003	**73.8**	64.8	7.9	56.9	6.9		28.2
2004	**73.7**	65.7	7.9	57.9	6.6		27.7
35–39							
1994	**20.2**	58.3	10.5	47.8	8.7		33.0
1995	**21.8**	59.6	10.4	49.2	8.7		31.7
1996	**24.7**	59.9	10.5	49.4	8.1		32.0
1997	**26.6**	59.8	10.0	49.8	8.0		32.2
1998	**28.7**	59.2	9.5	49.7	7.6		33.3
1999	**30.2**	60.0	9.5	50.6	7.0		33.0
2000	**31.7**	61.3	8.7	52.6	6.2		32.5
2001	**33.9**	61.6	8.6	53.0	6.4		32.1
2002	**37.5**	61.8	8.2	53.6	6.4		31.9
2003	**40.2**	62.0	8.2	53.8	6.2		31.8
2004	**41.7**	62.6	7.9	54.7	5.9		31.5

1 Conceptions leading to births outside marriage registered by the mother alone.
2 Conceptions leading to births outside marriage registered by both parents.

Table 12.4 - *continued*

Age of woman at conception and year of conception	Total number of conceptions outside marriage (000s)	Percentage of conceptions outside marriage					Age of woman at conception and year of conception	Total number of conceptions outside marriage (000s)	Percentage of conceptions outside marriage				
		Leading to maternities				Termi-nated by abortion			Leading to maternities				Termi-nated by abortion
		Outside marriage			Within marriage				Outside marriage			Within marriage	
		Total	Sole[1]	Joint[2]					Total	Sole[1]	Joint[2]		
40 and over													
1994	**4.3**	44.8	8.2	36.5	6.6	48.7							
1995	**4.7**	47.4	9.1	38.3	7.0	45.7							
1996	**5.2**	48.1	8.9	39.2	6.6	45.3							
1997	**5.5**	48.7	8.9	39.8	6.2	45.0							
1998	**5.9**	49.2	9.1	40.2	6.0	44.7							
1999	**6.3**	50.5	8.1	42.4	5.6	43.9							
2000	**6.7**	51.1	8.0	43.1	5.6	43.4							
2001	**7.4**	51.9	8.3	43.6	5.6	42.5							
2002	**8.5**	50.5	7.8	42.7	5.3	44.2							
2003	**9.5**	49.6	8.1	41.5	4.8	45.6							
2004	**10.5**	51.1	7.5	43.6	5.0	43.9							

1 Conceptions leading to births outside marriage registered by the mother alone.
2 Conceptions leading to births outside marriage registered by both parents.

Table 12.5 Conceptions within marriage (numbers and rates) : outcome and age of woman at conception, 1994-2004

England and Wales
Residents

Age of woman at conception and year of conception		Numbers of conceptions (thousands)			Conception rates per 1,000 married women in age-group[4]		
		Total conceptions within marriage	Conceptions leading to maternities within marriage	Conceptions terminated by abortion	**Total conceptions within marriage**	Conceptions leading to maternities within marriage	Conceptions terminated by abortion
All ages[1]	1994	**435.9**	401.6	34.3	**82.1**	75.6	6.5
	1995	**417.7**	384.7	33.1	**80.7**	74.3	6.4
	1996	**416.3**	381.8	34.5	**82.1**	75.3	6.8
	1997	**401.5**	367.6	34.0	**80.9**	74.1	6.8
	1998	**388.8**	354.6	34.2	**80.0**	73.0	7.0
	1999	**373.5**	340.8	32.7	**78.3**	71.4	6.9
	2000	**366.2**	335.0	31.2	**78.0**	71.3	6.6
	2001	**359.2**	328.3	30.9	**77.9**	71.2	6.7
	2002	**365.7**	337.0	28.7	**81.4**	75.0	6.4
	2003	**367.0**	340.4	26.6	**83.9**	77.9	6.1
	2004	**372.7**	346.6	26.1	**87.4**	81.3	6.1
Under 16	1994	**0.0**	0.0	0.0	:	:	:
	1995	**0.0**	0.0	0.0	:	:	:
	1996	**0.0**	0.0	0.0	:	:	:
	1997	**0.0**	0.0	0.0	:	:	:
	1998	**0.0**	0.0	0.0	:	:	:
	1999	**0.0**	0.0	0.0	:	:	:
	2000	**0.0**	0.0	0.0	:	:	:
	2001	**0.0**	0.0	0.0	:	:	:
	2002	**0.0**	0.0	0.0	:	:	:
	2003	**0.0**	0.0	0.0	:	:	:
	2004	**0.0**	0.0	0.0	:	:	:
16	1994	**0.2**	0.2	0.0	:	:	:
	1995	**0.2**	0.2	0.0	:	:	:
	1996	**0.3**	0.2	0.0	:	:	:
	1997	**0.3**	0.2	0.0	:	:	:
	1998	**0.2**	0.2	0.0	:	:	:
	1999	**0.2**	0.2	0.0	:	:	:
	2000	**0.2**	0.2	0.0	:	:	:
	2001	**0.2**	0.2	0.0	:	:	:
	2002	**0.2**	0.2	0.0	:	:	:
	2003	**0.2**	0.2	0.0	:	:	:
	2004	**0.2**	0.1	0.0	:	:	:
17	1994	**0.8**	0.8	0.1	:	:	:
	1995	**0.7**	0.7	0.0	:	:	:
	1996	**0.7**	0.7	0.0	:	:	:
	1997	**0.8**	0.7	0.1	:	:	:
	1998	**0.8**	0.7	0.1	:	:	:
	1999	**0.7**	0.6	0.1	:	:	:
	2000	**0.7**	0.6	0.1	:	:	:
	2001	**0.7**	0.6	0.0	:	:	:
	2002	**0.6**	0.6	0.1	:	:	:
	2003	**0.6**	0.6	0.0	:	:	:
	2004	**0.5**	0.5	0.0	:	:	:
18	1994	**2.0**	1.9	0.1	**377.2**	359.8	17.4
	1995	**2.0**	1.9	0.1	**384.5**	364.9	19.6
	1996	**1.8**	1.7	0.1	**356.7**	336.9	19.9
	1997	**1.9**	1.8	0.1	**368.6**	347.8	20.7
	1998	**2.0**	1.9	0.1	**380.2**	355.1	25.1
	1999	**1.9**	1.8	0.1	**386.7**	358.4	28.3
	2000	**1.9**	1.8	0.1	**453.2**	423.7	29.5
	2001	**1.8**	1.7	0.1	**414.8**	388.3	26.5
	2002	**1.8**	1.7	0.1	**495.7**	460.0	35.7
	2003	**1.5**	1.4	0.1	**456.3**	425.5	30.8
	2004	**1.5**	1.4	0.1	**619.8**	574.1	45.7

Note: Rates for ages under 16, 16 and 17 have been omitted because of the small number of married women at these ages.
1 Rates per 1,000 women aged 15-44.
4 See Section 2.1.

Table 12.5 - *continued*

Age of woman at conception and year of conception		Numbers of conceptions (thousands)			Conception rates per 1,000 married women in age-group[4]		
		Total conceptions within marriage	Conceptions leading to maternities within marriage	Conceptions terminated by abortion	Total conceptions within marriage	Conceptions leading to maternities within marriage	Conceptions terminated by abortion
19	1994	**4.0**	3.8	0.2	**265.3**	251.6	13.7
	1995	**3.7**	3.5	0.2	**258.3**	245.4	12.9
	1996	**3.7**	3.4	0.2	**259.0**	244.3	14.7
	1997	**3.5**	3.2	0.2	**254.9**	239.7	15.3
	1998	**3.7**	3.5	0.2	**275.8**	258.6	17.3
	1999	**3.6**	3.3	0.2	**279.8**	261.2	18.6
	2000	**3.7**	3.4	0.2	**334.6**	312.4	22.2
	2001	**3.6**	3.4	0.2	**398.7**	375.2	23.5
	2002	**3.4**	3.2	0.2	**434.3**	406.3	28.0
	2003	**3.2**	3.0	0.2	**431.2**	403.4	27.8
	2004	**3.2**	3.0	0.2	**439.0**	411.6	27.4
Under 20[2]	1994	**7.0**	6.6	0.4	**325.3**	308.5	16.8
	1995	**6.6**	6.3	0.3	**318.4**	301.8	16.6
	1996	**6.5**	6.1	0.4	**313.2**	295.4	17.8
	1997	**6.4**	6.0	0.4	**315.9**	296.8	19.1
	1998	**6.7**	6.3	0.5	**332.4**	309.9	22.4
	1999	**6.4**	6.0	0.5	**325.8**	302.9	22.9
	2000	**6.5**	6.0	0.4	**367.6**	342.8	24.8
	2001	**6.3**	5.9	0.4	**387.8**	364.5	23.3
	2002	**6.1**	5.7	0.4	**468.2**	435.4	32.8
	2003	**5.6**	5.2	0.4	**476.0**	444.0	32.0
	2004	**5.5**	5.1	0.4	**517.0**	480.9	36.2
20–24	1994	**71.8**	67.9	3.9	**208.0**	196.7	11.3
	1995	**64.0**	60.4	3.6	**213.3**	201.4	11.9
	1996	**58.4**	54.9	3.5	**224.8**	211.3	13.5
	1997	**52.3**	49.1	3.2	**232.5**	218.1	14.4
	1998	**48.9**	45.7	3.2	**243.1**	227.0	16.1
	1999	**45.6**	42.5	3.1	**242.9**	226.1	16.7
	2000	**45.0**	41.8	3.2	**250.8**	232.9	17.9
	2001	**45.1**	41.8	3.3	**253.2**	234.6	18.6
	2002	**45.4**	42.2	3.2	**273.4**	254.0	19.4
	2003	**45.7**	42.5	3.1	**283.9**	264.6	19.3
	2004	**46.2**	43.2	3.1	**295.7**	276.1	19.6
25–29	1994	**169.2**	160.1	9.1	**165.5**	156.6	8.9
	1995	**156.8**	148.1	8.7	**162.9**	153.9	9.0
	1996	**152.6**	143.9	8.7	**168.4**	158.8	9.5
	1997	**142.9**	134.5	8.5	**169.3**	159.3	10.0
	1998	**133.2**	125.2	8.0	**170.2**	160.0	10.2
	1999	**123.5**	116.1	7.5	**170.4**	160.1	10.3
	2000	**116.9**	109.8	7.1	**172.7**	162.2	10.5
	2001	**108.9**	102.1	6.9	**174.2**	163.2	11.0
	2002	**108.1**	101.8	6.3	**190.6**	179.5	11.1
	2003	**106.1**	100.3	5.8	**202.7**	191.5	11.2
	2004	**107.8**	101.8	6.0	**217.0**	204.9	12.1
30–34	1994	**133.5**	122.9	10.5	**99.9**	92.0	7.9
	1995	**134.9**	124.6	10.3	**101.4**	93.7	7.8
	1996	**139.2**	128.3	10.9	**105.8**	97.5	8.3
	1997	**138.4**	127.6	10.7	**107.0**	98.7	8.3
	1998	**136.3**	125.3	11.0	**108.2**	99.5	8.7
	1999	**132.4**	122.2	10.2	**108.3**	100.0	8.3
	2000	**130.4**	121.0	9.5	**110.3**	102.3	8.0
	2001	**130.1**	120.8	9.4	**113.9**	105.7	8.2
	2002	**133.6**	125.2	8.4	**122.1**	114.4	7.7
	2003	**135.3**	127.6	7.6	**129.7**	122.4	7.3
	2004	**136.0**	128.6	7.3	**138.1**	130.6	7.4

Note: Rates for ages under 16, 16 and 17 have been omitted because of the small number of married women at these ages.
2 Rates per 1,000 women aged 15-19.
4 See Section 2.1.

Table 12.5 - *continued*

Age of woman at conception and year of conception		Numbers of conceptions (thousands)			Conception rates per 1,000 married women in age-group[4]		
		Total conceptions within marriage	Conceptions leading to maternities within marriage	Conceptions terminated by abortion	**Total conceptions within marriage**	Conceptions leading to maternities within marriage	Conceptions terminated by abortion
35–39	1994	**46.0**	38.7	7.3	**35.5**	29.8	5.6
	1995	**46.9**	39.6	7.3	**35.9**	30.3	5.6
	1996	**50.8**	42.7	8.1	**38.4**	32.3	6.1
	1997	**52.3**	44.2	8.0	**39.2**	33.2	6.0
	1998	**54.1**	45.8	8.3	**40.3**	34.2	6.2
	1999	**55.8**	47.5	8.3	**41.2**	35.1	6.1
	2000	**57.0**	49.1	7.9	**41.8**	36.1	5.8
	2001	**58.4**	50.4	8.0	**43.1**	37.1	5.9
	2002	**61.4**	54.1	7.3	**45.8**	40.3	5.5
	2003	**62.9**	56.2	6.7	**47.9**	42.8	5.1
	2004	**65.0**	58.6	6.4	**50.6**	45.6	5.0
40 and over[3]	1994	**8.5**	5.4	3.2	**6.6**	4.2	2.4
	1995	**8.5**	5.7	2.9	**6.8**	4.5	2.3
	1996	**8.9**	6.0	3.0	**7.1**	4.8	2.4
	1997	**9.2**	6.1	3.1	**7.4**	4.9	2.5
	1998	**9.5**	6.3	3.2	**7.6**	5.0	2.6
	1999	**9.7**	6.6	3.2	**7.7**	5.2	2.5
	2000	**10.3**	7.2	3.1	**8.1**	5.6	2.4
	2001	**10.5**	7.4	3.0	**8.1**	5.7	2.3
	2002	**11.0**	8.0	3.0	**8.4**	6.1	2.3
	2003	**11.5**	8.5	3.0	**8.7**	6.4	2.2
	2004	**12.2**	9.4	2.9	**9.2**	7.0	2.2

Note: Rates for ages under 16, 16 and 17 have been omitted because of the small number of married women at these ages.

3 Rates per 1,000 women aged 40–44.

4 See Section 2.1.

Table 12.6   Conceptions outside marriage (numbers and rates) : outcome and age of woman at conception, 1994-2004   **England and Wales** *Residents*

Age of woman at conception and year of conception		Numbers of conceptions (thousands)					Conception rates per 1,000 unmarried women in age-group[5]				
		Total conceptions outside marriage	Conceptions leading to maternities			Conceptions terminated by abortion	**Total conceptions outside marriage**	Conceptions leading to maternities			Conceptions terminated by abortion
			Total	Outside marriage	Within marriage following marriage after conception			Total	Outside marriage	Within marriage following marriage after conception	
All ages[1]	1994	**365.6**	244.0	217.7	26.3	121.7	**67.6**	45.1	40.2	4.9	22.5
	1995	**372.5**	249.8	224.2	25.6	122.7	**67.3**	45.1	40.5	4.6	22.2
	1996	**400.6**	265.4	239.7	25.6	135.3	**70.9**	47.0	42.5	4.5	24.0
	1997	**398.9**	262.3	237.3	25.0	136.6	**69.2**	45.5	41.2	4.3	23.7
	1998	**408.2**	265.0	241.0	24.0	143.2	**69.4**	45.1	41.0	4.1	24.4
	1999	**400.4**	258.5	236.5	22.0	141.9	**66.8**	43.1	39.5	3.7	23.7
	2000	**400.8**	257.6	236.7	20.8	143.2	**65.4**	42.1	38.7	3.4	23.4
	2001	**404.4**	258.3	238.2	20.1	146.1	**64.6**	41.3	38.1	3.2	23.4
	2002	**421.3**	272.8	252.0	20.9	148.5	**65.7**	42.5	39.3	3.3	23.1
	2003	**439.8**	285.0	263.5	21.5	154.8	**66.9**	43.4	40.1	3.3	23.6
	2004	**454.1**	295.1	274.4	20.7	159.0	**67.6**	43.9	40.8	3.1	23.7
Under 16[2]	1994	**7.8**	3.8	3.8	0.0	3.9	**8.3**	4.1	4.1	0.0	4.2
	1995	**8.0**	4.2	4.1	0.1	3.8	**8.5**	4.5	4.4	0.1	4.1
	1996	**8.8**	4.5	4.4	0.1	4.4	**9.5**	4.8	4.7	0.1	4.7
	1997	**8.2**	4.1	4.1	0.0	4.1	**8.9**	4.5	4.4	0.0	4.4
	1998	**8.4**	4.0	4.0	0.0	4.4	**8.9**	4.2	4.2	0.0	4.7
	1999	**7.9**	3.7	3.7	0.0	4.2	**8.3**	3.9	3.9	0.0	4.4
	2000	**8.1**	3.7	3.7	0.0	4.4	**8.3**	3.8	3.8	0.0	4.5
	2001	**7.9**	3.5	3.4	0.0	4.4	**8.0**	3.5	3.5	0.0	4.5
	2002	**7.8**	3.5	3.4	0.0	4.4	**7.8**	3.5	3.4	0.0	4.4
	2003	**8.0**	3.4	3.4	0.0	4.6	**7.9**	3.4	3.3	0.0	4.6
	2004	**7.6**	3.2	3.2	0.0	4.3	**7.5**	3.2	3.2	0.0	4.3
16	1994	**11.1**	6.6	6.4	0.2	4.5	**39.6**	23.5	22.9	0.7	16.0
	1995	**12.1**	7.4	7.3	0.2	4.7	**39.8**	24.4	23.8	0.6	15.4
	1996	**14.0**	8.4	8.2	0.2	5.7	**44.2**	26.4	25.8	0.5	17.9
	1997	**13.8**	8.1	8.0	0.2	5.7	**43.7**	25.7	25.2	0.5	18.0
	1998	**13.6**	7.9	7.7	0.1	5.7	**43.9**	25.4	25.0	0.5	18.4
	1999	**13.1**	7.3	7.2	0.1	5.8	**42.5**	23.6	23.2	0.4	18.9
	2000	**13.0**	7.1	7.0	0.1	5.8	**41.8**	22.9	22.5	0.4	18.8
	2001	**12.9**	6.9	6.8	0.1	6.0	**39.8**	21.3	20.9	0.4	18.5
	2002	**13.2**	7.1	7.0	0.1	6.2	**40.8**	21.8	21.4	0.3	19.0
	2003	**13.1**	7.0	6.9	0.1	6.1	**39.5**	21.2	20.9	0.3	18.3
	2004	**13.5**	7.3	7.2	0.1	6.2	**39.5**	21.3	21.1	0.3	18.2
17	1994	**16.1**	10.3	9.8	0.5	5.9	**58.7**	37.4	35.5	1.9	21.4
	1995	**16.7**	10.7	10.2	0.5	6.1	**59.8**	38.1	36.4	1.6	21.7
	1996	**19.6**	12.3	11.8	0.5	7.3	**64.5**	40.4	38.8	1.6	24.1
	1997	**20.3**	12.5	12.1	0.5	7.7	**63.9**	39.5	38.0	1.5	24.4
	1998	**21.1**	12.8	12.3	0.5	8.3	**66.9**	40.6	39.1	1.5	26.3
	1999	**20.0**	12.0	11.7	0.4	8.0	**64.9**	39.0	37.7	1.2	25.9
	2000	**19.4**	11.4	11.1	0.3	8.0	**63.2**	37.2	36.1	1.1	25.9
	2001	**19.3**	11.0	10.7	0.3	8.3	**62.1**	35.4	34.3	1.1	26.7
	2002	**20.0**	11.6	11.3	0.3	8.4	**61.4**	35.6	34.6	1.0	25.8
	2003	**20.2**	11.7	11.5	0.3	8.5	**61.9**	35.9	35.0	0.8	26.0
	2004	**20.4**	11.8	11.5	0.3	8.6	**61.1**	35.3	34.5	0.8	25.8

1 Rates per 1,000 women aged 15-44.
2 Rates per 1,000 women aged 13-15.
5 See Section 2.1.

Table 12.6 - *continued*

Age of woman at conception and year of conception		Numbers of conceptions (thousands)					Conception rates per 1,000 unmarried women in age-group[5]				
		Total con-ceptions outside marriage	Conceptions leading to maternities			Conceptions terminated by abortion	Total con-ceptions outside marriage	Conceptions leading to maternities			Conceptions terminated by abortion
			Total	Outside marriage	Within marriage following marriage after conception			Total	Outside marriage	Within marriage following marriage after conception	
18	1994	**20.7**	13.5	12.6	0.9	7.1	**73.1**	47.8	44.6	3.2	25.2
	1995	**20.4**	13.2	12.4	0.8	7.2	**75.4**	48.7	45.7	3.0	26.7
	1996	**22.3**	14.2	13.3	0.9	8.1	**81.0**	51.4	48.3	3.1	29.5
	1997	**23.7**	15.0	14.2	0.8	8.7	**79.2**	50.2	47.6	2.6	29.0
	1998	**25.9**	16.1	15.3	0.8	9.8	**83.3**	51.6	49.1	2.6	31.6
	1999	**24.7**	15.2	14.5	0.7	9.5	**79.7**	49.1	46.7	2.4	30.6
	2000	**24.3**	14.7	14.1	0.6	9.5	**80.7**	49.0	47.0	2.0	31.7
	2001	**24.0**	14.4	13.8	0.6	9.6	**79.0**	47.5	45.5	2.0	31.5
	2002	**24.1**	14.7	14.1	0.6	9.5	**77.1**	46.9	45.0	1.8	30.3
	2003	**25.1**	15.1	14.6	0.6	9.9	**76.7**	46.3	44.5	1.8	30.4
	2004	**25.8**	15.6	15.0	0.6	10.2	**78.4**	47.4	45.7	1.7	31.1
19	1994	**22.7**	14.8	13.6	1.2	7.9	**78.2**	51.1	47.0	4.1	27.1
	1995	**22.6**	14.9	13.8	1.1	7.8	**80.9**	53.1	49.3	3.9	27.7
	1996	**23.6**	15.1	14.0	1.1	8.5	**87.9**	56.1	52.2	4.0	31.8
	1997	**23.6**	14.9	13.9	0.9	8.7	**86.5**	54.5	51.1	3.4	32.0
	1998	**25.9**	16.2	15.2	1.0	9.7	**87.7**	54.9	51.5	3.3	32.9
	1999	**26.5**	16.4	15.5	0.9	10.2	**86.4**	53.3	50.3	3.0	33.1
	2000	**26.5**	16.2	15.3	0.9	10.3	**86.8**	53.1	50.3	2.8	33.7
	2001	**25.7**	15.5	14.7	0.8	10.1	**85.4**	51.7	49.0	2.7	33.6
	2002	**25.9**	15.9	15.2	0.8	9.9	**84.9**	52.3	49.8	2.5	32.7
	2003	**26.6**	16.4	15.7	0.8	10.2	**84.6**	52.2	49.8	2.4	32.4
	2004	**28.5**	17.7	17.0	0.7	10.8	**86.7**	53.8	51.6	2.2	32.9
Under 20[3]	1994	**78.3**	49.1	46.2	2.8	29.3	**54.7**	34.3	32.3	2.0	20.4
	1995	**80.0**	50.4	47.8	2.6	29.6	**55.0**	34.7	32.9	1.8	20.4
	1996	**88.4**	54.4	51.7	2.6	34.0	**59.7**	36.7	34.9	1.8	23.0
	1997	**89.6**	54.7	52.3	2.4	34.9	**59.2**	36.1	34.5	1.6	23.1
	1998	**94.9**	56.9	54.5	2.5	38.0	**61.6**	37.0	35.4	1.6	24.6
	1999	**92.4**	54.7	52.5	2.2	37.7	**59.7**	35.3	33.9	1.4	24.4
	2000	**91.2**	53.2	51.2	2.0	38.0	**59.0**	34.4	33.1	1.3	24.6
	2001	**89.7**	51.3	49.4	1.9	38.4	**57.4**	32.9	31.6	1.2	24.6
	2002	**91.0**	52.7	50.9	1.8	38.3	**57.0**	33.0	31.9	1.1	24.0
	2003	**93.0**	53.7	52.0	1.7	39.3	**56.8**	32.8	31.7	1.1	24.0
	2004	**95.8**	55.6	53.9	1.7	40.2	**57.4**	33.3	32.3	1.0	24.1
20–24	1994	**118.6**	77.9	69.5	8.4	40.7	**83.0**	54.5	48.6	5.9	28.5
	1995	**117.1**	76.9	69.1	7.9	40.2	**83.0**	54.5	48.9	5.6	28.5
	1996	**121.4**	78.7	71.2	7.4	42.8	**88.4**	57.3	51.9	5.4	31.2
	1997	**115.0**	73.6	66.8	6.7	41.4	**86.1**	55.1	50.0	5.0	31.0
	1998	**114.4**	72.2	65.9	6.3	42.2	**86.9**	54.9	50.1	4.8	32.1
	1999	**112.0**	70.2	64.5	5.7	41.8	**84.3**	52.8	48.5	4.3	31.5
	2000	**113.9**	70.7	65.2	5.5	43.2	**83.7**	52.0	47.9	4.0	31.8
	2001	**116.6**	71.8	66.7	5.1	44.8	**83.3**	51.3	47.7	3.7	32.0
	2002	**122.3**	77.3	71.9	5.4	45.1	**85.1**	53.7	50.0	3.7	31.3
	2003	**129.6**	82.0	76.4	5.6	47.6	**87.8**	55.6	51.7	3.8	32.3
	2004	**135.1**	85.7	80.4	5.3	49.4	**89.5**	56.8	53.3	3.5	32.7

3 Rates per 1,000 women aged 15-19.
5 See Section 2.1.

Table 12.6 - *continued*

Age of woman at conception and year of conception		Numbers of conceptions (thousands)					Conception rates per 1,000 unmarried women in age-group[5]				
		Total conceptions outside marriage	Conceptions leading to maternities			Conceptions terminated by abortion	Total conceptions outside marriage	Conceptions leading to maternities			Conceptions terminated by abortion
			Total	Outside marriage	Within marriage following marriage after conception			Total	Outside marriage	Within marriage following marriage after conception	
25–29	1994	**92.6**	64.2	56.2	8.1	28.4	**90.9**	63.0	55.1	7.9	27.9
	1995	**93.5**	65.1	57.1	8.0	28.4	**89.7**	62.4	54.8	7.6	27.3
	1996	**100.0**	69.2	61.2	7.9	30.9	**93.2**	64.4	57.1	7.4	28.8
	1997	**99.6**	68.3	60.3	7.9	31.3	**90.5**	62.0	54.8	7.2	28.5
	1998	**99.2**	67.4	59.9	7.5	31.8	**88.6**	60.2	53.5	6.7	28.4
	1999	**94.9**	64.1	57.4	6.8	30.8	**84.3**	57.0	51.0	6.0	27.4
	2000	**92.4**	62.5	56.2	6.3	29.9	**81.6**	55.2	49.6	5.6	26.4
	2001	**90.4**	60.6	54.7	5.9	29.7	**80.7**	54.2	48.9	5.3	26.6
	2002	**91.3**	61.9	56.0	5.9	29.3	**82.5**	56.0	50.6	5.3	26.5
	2003	**93.7**	63.8	57.7	6.1	29.9	**84.9**	57.8	52.3	5.5	27.1
	2004	**97.4**	66.1	60.1	5.9	31.3	**86.2**	58.5	53.3	5.2	27.7
30–34	1994	**51.5**	37.0	32.1	4.9	14.6	**76.7**	55.0	47.7	7.3	21.7
	1995	**55.4**	39.9	34.9	5.0	15.5	**76.9**	55.4	48.5	6.9	21.4
	1996	**60.8**	43.5	38.2	5.3	17.3	**80.0**	57.2	50.3	7.0	22.8
	1997	**62.6**	44.7	39.2	5.5	17.9	**78.7**	56.2	49.4	6.9	22.5
	1998	**65.1**	46.1	40.9	5.2	19.0	**79.2**	56.1	49.8	6.3	23.2
	1999	**64.7**	45.8	40.8	5.0	18.9	**76.4**	54.1	48.2	5.9	22.3
	2000	**64.9**	46.0	41.3	4.7	18.9	**74.8**	53.0	47.6	5.4	21.8
	2001	**66.5**	47.2	42.6	4.6	19.3	**74.7**	53.0	47.8	5.2	21.7
	2002	**70.6**	50.7	45.7	5.0	20.0	**77.1**	55.3	49.9	5.4	21.8
	2003	**73.8**	53.0	47.8	5.1	20.8	**79.2**	56.8	51.3	5.5	22.3
	2004	**73.7**	53.3	48.4	4.9	20.4	**79.1**	57.2	51.9	5.2	21.9
35–39	1994	**20.2**	13.5	11.8	1.8	6.7	**43.6**	29.2	25.4	3.8	14.4
	1995	**21.8**	14.9	13.0	1.9	6.9	**43.9**	30.0	26.2	3.8	13.9
	1996	**24.7**	16.8	14.8	2.0	7.9	**46.3**	31.5	27.8	3.8	14.8
	1997	**26.6**	18.0	15.9	2.1	8.6	**46.5**	31.6	27.8	3.7	15.0
	1998	**28.7**	19.2	17.0	2.2	9.6	**47.1**	31.4	27.9	3.6	15.7
	1999	**30.2**	20.2	18.1	2.1	10.0	**46.4**	31.1	27.9	3.2	15.3
	2000	**31.7**	21.4	19.4	2.0	10.3	**45.8**	31.0	28.1	2.9	14.9
	2001	**33.9**	23.0	20.8	2.2	10.9	**46.6**	31.7	28.7	3.0	15.0
	2002	**37.5**	25.5	23.1	2.4	11.9	**49.3**	33.6	30.4	3.1	15.7
	2003	**40.2**	27.4	24.9	2.5	12.8	**51.1**	34.9	31.7	3.2	16.2
	2004	**41.7**	28.6	26.1	2.5	13.1	**51.6**	35.4	32.3	3.1	16.3
40 and over[4]	1994	**4.3**	2.2	1.9	0.3	2.1	**10.9**	5.6	4.9	0.7	5.3
	1995	**4.7**	2.6	2.2	0.3	2.2	**11.5**	6.3	5.5	0.8	5.3
	1996	**5.2**	2.8	2.5	0.3	2.4	**12.2**	6.7	5.9	0.8	5.5
	1997	**5.5**	3.0	2.7	0.3	2.5	**12.3**	6.7	6.0	0.8	5.5
	1998	**5.9**	3.2	2.9	0.4	2.6	**12.5**	6.9	6.1	0.7	5.6
	1999	**6.3**	3.6	3.2	0.4	2.8	**12.7**	7.2	6.4	0.7	5.6
	2000	**6.7**	3.8	3.4	0.4	2.9	**12.7**	7.2	6.5	0.7	5.5
	2001	**7.4**	4.2	3.8	0.4	3.1	**13.2**	7.6	6.8	0.7	5.6
	2002	**8.5**	4.8	4.3	0.5	3.8	**14.3**	8.0	7.2	0.8	6.3
	2003	**9.5**	5.2	4.7	0.5	4.3	**14.9**	8.1	7.4	0.7	6.8
	2004	**10.5**	5.9	5.4	0.5	4.6	**15.6**	8.7	8.0	0.8	6.8

4   Rates per 1,000 women aged 40–44.
5   See Section 2.1.

Table 12.7 Conceptions (numbers and percentages): occurrence within/outside marriage and outcome, area of usual residence, and age of woman at conception, 2004

England and Wales, Government Office Regions (within England), counties, unitary authorities and London boroughs

Residents

Area of usual residence	Total number of conceptions to all women (000s)	Percentage of all conceptions		Total number of conceptions outside marriage (000s)	Percentage of conceptions outside marriage					Total number of conceptions to women aged under 18	Percentage of under 18 conceptions terminated by abortion
		Outside marriage	Terminated by abortion		Leading to maternities				Terminated by abortion		
					Outside marriage			Within marriage following marriage after conception			
					Total	Sole[1]	Joint[2]				
England and Wales	**826.8**	**54.9**	**22.4**	**454.1**	**60.4**	**10.0**	**50.5**	**4.6**	**35.0**	**42,198**	**45.6**
England	786.5	54.6	22.6	429.2	59.9	9.9	50.0	4.6	35.5	39,593	46.0
North East	**35.0**	**64.5**	**20.3**	**22.6**	**68.0**	**11.2**	**56.8**	**3.6**	**28.5**	**2,515**	**40.2**
Darlington UA	**1.5**	61.9	17.2	**0.9**	70.3	12.2	58.1	4.6	25.1	93	41.9
Hartlepool UA	**1.4**	71.8	22.4	**1.0**	68.6	9.6	58.9	2.7	28.7	126	42.1
Middlesbrough UA	**2.4**	65.6	20.3	**1.6**	69.3	16.9	52.4	2.6	28.1	191	37.7
Redcar and Cleveland UA	**1.8**	70.3	19.6	**1.3**	71.6	12.4	59.2	3.2	25.2	173	33.5
Stockton-on-Tees UA	**2.8**	63.2	20.3	**1.7**	67.9	11.5	56.4	4.1	28.0	185	38.9
Durham	6.3	64.2	17.7	4.1	71.2	9.7	61.5	3.8	25.0	445	38.9
Northumberland	3.6	59.5	19.8	2.2	66.1	8.5	57.7	4.4	29.5	214	44.9
Tyne and Wear	15.2	64.9	21.6	9.8	66.0	11.3	54.7	3.4	30.6	1,088	41.1
North West	**102.2**	**59.7**	**20.2**	**61.0**	**66.0**	**11.8**	**54.1**	**3.5**	**30.5**	**6,268**	**42.5**
Blackburn with Darwin UA	**2.8**	47.5	15.6	**1.3**	68.3	11.0	57.3	5.6	26.2	179	38.5
Blackpool UA	**2.1**	72.7	21.8	**1.5**	69.3	11.0	58.3	3.7	26.9	193	31.6
Halton UA	**2.0**	68.9	21.2	**1.4**	69.2	13.2	56.0	3.1	27.7	115	40.9
Warrington UA	**2.8**	55.7	20.3	**1.5**	65.5	8.9	56.7	2.9	31.5	153	43.1
Cheshire	9.0	51.2	17.7	4.6	64.9	9.4	55.5	4.4	30.6	425	48.5
Cumbria	5.9	57.5	18.1	3.4	67.9	8.1	59.8	4.4	27.7	333	42.9
Greater Manchester	42.4	59.8	21.3	25.4	64.3	11.6	52.7	3.3	32.3	2,678	40.3
Lancashire	15.9	55.2	17.7	8.8	67.5	10.5	57.0	4.2	28.3	968	40.2
Merseyside	19.4	67.4	22.3	13.1	67.1	15.4	51.7	2.6	30.3	1,224	49.3
Yorkshire and the Humber	**75.0**	**57.1**	**19.3**	**42.8**	**65.6**	**10.5**	**55.1**	**4.2**	**30.2**	**4,698**	**38.8**
East Riding of Yorkshire UA	**3.5**	51.8	17.0	**1.8**	66.7	7.4	59.3	5.4	27.9	190	45.8
Kingston upon Hull, City of UA	**4.1**	73.1	20.5	**3.0**	70.8	11.3	59.5	3.7	25.5	412	32.3
North East Lincolnshire UA	**2.5**	72.3	21.7	**1.8**	67.1	11.4	55.6	3.0	29.9	239	40.2
North Lincolnshire UA	**2.2**	61.5	18.6	**1.3**	68.8	11.1	57.7	4.0	27.2	156	43.6
York UA	**2.4**	55.0	20.9	**1.3**	59.9	7.2	52.7	5.4	34.7	113	47.8
North Yorkshire	6.9	48.5	16.9	3.3	64.1	8.2	55.9	6.6	29.3	290	46.9
South Yorkshire	19.1	61.9	20.9	11.8	66.2	10.3	55.9	3.7	30.1	1,351	39.7
West Yorkshire	34.4	53.5	18.7	18.4	64.7	11.4	53.3	4.1	31.3	1,947	36.7
East Midlands	**60.5**	**55.8**	**19.7**	**33.8**	**65.3**	**9.7**	**55.6**	**4.6**	**30.1**	**3,368**	**40.9**
Derby UA	**3.8**	57.5	22.1	**2.2**	62.1	11.2	50.9	4.1	33.8	241	32.4
Leicester UA & Rutland UA	**6.4**	51.6	22.8	**3.3**	60.0	12.1	47.9	4.7	35.3	300	35.7
Nottingham UA	**4.9**	68.5	26.3	**3.4**	62.7	15.5	47.2	3.0	34.4	376	31.6
Derbyshire	9.4	55.7	16.5	5.3	69.4	8.0	61.4	5.2	25.3	506	45.1
Leicestershire	8.0	49.7	18.0	4.0	65.4	6.6	58.8	4.6	30.0	353	51.3
Lincolnshire	8.0	56.9	18.3	4.6	67.4	9.0	58.4	5.2	27.4	503	42.5
Northamptonshire	10.2	55.1	19.8	5.6	64.4	9.1	55.2	4.7	31.0	558	43.0
Nottinghamshire	9.7	56.6	18.8	5.5	66.6	9.0	57.5	4.7	28.7	531	39.9

1 Conceptions leading to births outside marriage registered by the mother alone.
2 Conceptions leading to births outside marriage registered by both parents.

Table 12.7 - *continued*

Area of usual residence	Total number of conceptions to all women (000s)	Percentage of all conceptions		Total number of conceptions outside marriage (000s)	Percentage of conceptions outside marriage					Total number of conceptions to women aged under 18	Percentage of under 18 conceptions terminated by abortion
		Outside marriage	Terminated by abortion		Leading to maternities				Terminated by abortion		
					Outside marriage			Within marriage following marriage after conception			
					Total	Sole[1]	Joint[2]				
West Midlands	**84.4**	54.7	22.0	**46.1**	62.1	10.9	51.2	4.1	33.8	**4,795**	43.2
Herefordshire, County of UA	**2.0**	53.3	17.4	**1.1**	64.8	8.5	56.3	5.6	29.6	**126**	46.0
Stoke-on-Trent UA	**4.1**	64.5	17.8	**2.6**	71.8	12.3	59.5	3.7	24.5	**320**	32.5
Telford and Wrekin UA	**2.6**	60.9	20.1	**1.6**	68.5	11.1	57.4	4.1	27.5	**182**	26.9
Shropshire	**3.4**	51.9	17.0	**1.8**	65.5	6.3	59.2	5.9	28.6	**148**	49.3
Staffordshire	**10.6**	55.0	20.0	**5.8**	64.8	8.3	56.5	4.9	30.3	**550**	47.5
Warwickshire	**7.1**	52.4	21.7	**3.7**	60.1	8.9	51.3	5.0	34.9	**371**	47.7
West Midlands	**47.2**	54.4	24.2	**25.7**	59.7	12.4	47.3	3.5	36.9	**2,752**	43.5
Worcestershire	**7.4**	51.6	18.1	**3.8**	65.1	9.0	56.1	5.3	29.6	**346**	43.9
East	**79.5**	50.9	19.2	**40.5**	62.3	8.3	54.0	5.3	32.3	**3,392**	47.0
Luton UA	**4.2**	44.7	22.9	**1.9**	52.4	13.0	39.4	5.2	42.3	**182**	47.3
Peterborough UA	**3.0**	51.8	19.2	**1.6**	64.3	12.4	51.9	5.7	30.0	**175**	30.9
Southend-on-Sea UA	**2.5**	60.1	22.8	**1.5**	62.6	11.0	51.6	3.8	33.6	**135**	41.5
Thurrock UA	**2.8**	59.7	21.3	**1.7**	62.7	7.6	55.1	5.1	32.2	**123**	48.0
Bedfordshire	**6.0**	48.7	19.8	**2.9**	61.5	8.7	52.8	5.5	33.0	**242**	51.7
Cambridgeshire	**7.6**	45.5	16.3	**3.5**	64.8	7.4	57.4	6.0	29.3	**267**	44.9
Essex	**18.4**	52.7	20.2	**9.7**	60.8	7.7	53.1	4.8	34.4	**774**	52.1
Hertfordshire	**15.9**	47.0	19.3	**7.5**	60.0	7.2	52.8	4.9	35.2	**545**	50.5
Norfolk	**10.0**	56.9	18.7	**5.7**	66.6	8.0	58.6	5.6	27.8	**514**	43.0
Suffolk	**9.0**	50.7	16.7	**4.6**	66.2	8.6	57.6	6.7	27.1	**435**	44.6
London	**165.9**	52.6	30.8	**87.2**	45.7	10.1	35.6	5.0	49.3	**6,235**	59.1
Inner London	**73.7**	55.7	33.7	**41.0**	43.0	10.6	32.4	4.8	52.2	**2,705**	59.3
Camden	**4.4**	50.0	31.3	**2.2**	40.4	9.8	30.7	5.8	53.8	**127**	66.1
Hackney & City of London	**6.6**	58.7	32.9	**3.9**	45.1	10.4	34.7	4.7	50.2	**285**	59.3
Hammersmith and Fulham	**3.9**	53.3	32.4	**2.1**	41.4	11.0	30.4	5.0	53.6	**106**	52.8
Haringey	**5.9**	57.4	34.1	**3.4**	43.5	10.3	33.1	5.7	50.8	**283**	53.4
Islington	**4.3**	62.4	36.3	**2.7**	44.7	10.3	34.5	4.2	51.0	**156**	67.3
Kensington and Chelsea	**3.1**	46.3	31.9	**1.5**	34.2	6.9	27.3	6.7	59.0	**54**	63.0
Lambeth	**7.5**	65.9	37.8	**5.0**	44.6	11.8	32.8	4.1	51.3	**357**	61.9
Lewisham	**6.5**	65.4	36.2	**4.3**	47.0	11.5	35.5	4.1	48.9	**317**	57.4
Newham	**7.7**	49.2	31.7	**3.8**	43.4	13.3	30.1	5.2	51.4	**256**	60.2
Southwark	**7.5**	67.1	38.2	**5.0**	45.2	12.7	32.5	3.9	50.9	**343**	60.9
Tower Hamlets	**5.6**	41.3	30.0	**2.3**	38.0	8.4	29.6	5.5	56.5	**173**	57.8
Wandsworth	**6.3**	48.0	28.2	**3.0**	43.8	8.7	35.1	5.4	50.8	**174**	54.0
Westminster	**4.3**	46.2	34.3	**2.0**	32.6	6.2	26.4	5.4	62.0	**74**	62.2
Outer London	**92.2**	50.1	28.5	**46.2**	48.1	9.6	38.4	5.2	46.7	**3,530**	58.9
Barking and Dagenham	**4.2**	63.5	30.1	**2.7**	52.6	12.6	40.0	3.4	43.9	**211**	53.6
Barnet	**6.5**	43.6	28.3	**2.8**	39.9	8.0	31.9	7.3	52.8	**203**	68.5
Bexley	**3.6**	56.4	26.0	**2.0**	56.7	8.7	48.0	4.5	38.8	**172**	65.7
Brent	**6.9**	48.5	35.2	**3.3**	38.3	9.6	28.8	5.4	56.3	**266**	58.3
Bromley	**4.7**	52.3	23.6	**2.5**	57.3	6.8	50.5	4.9	37.7	**166**	60.2

1 Conceptions leading to births outside marriage registered by the mother alone.
2 Conceptions leading to births outside marriage registered by both parents.

Table 12.7 - *continued*

Area of usual residence	Total number of conceptions to all women (000s)	Percentage of all conceptions		Total number of conceptions outside marriage (000s)	Percentage of conceptions outside marriage					Total number of conceptions to women aged under 18	Percentage of under 18 conceptions terminated by abortion
		Outside marriage	Terminated by abortion		Leading to maternities				Terminated by abortion		
					Outside marriage			Within marriage following marriage after conception			
					Total	Sole[1]	Joint[2]				
Croydon	7.1	57.7	32.3	4.1	49.0	11.1	37.9	4.3	46.7	364	59.1
Ealing	6.8	43.6	28.9	3.0	42.7	8.5	34.2	6.1	51.2	193	57.5
Enfield	6.2	53.7	29.4	3.3	49.2	11.0	38.2	5.6	45.2	272	50.7
Greenwich	5.6	60.9	30.4	3.4	52.3	12.9	39.4	4.3	43.4	261	49.8
Harrow	4.1	37.0	29.7	1.5	38.2	9.0	29.2	4.7	57.1	141	67.4
Havering	3.4	59.1	27.7	2.0	53.2	7.2	46.1	3.9	42.8	152	60.5
Hillingdon	4.9	48.7	29.1	2.4	49.6	8.8	40.8	4.3	46.2	223	61.9
Hounslow	5.0	45.8	27.9	2.3	48.4	10.9	37.5	5.5	46.1	203	51.2
Kingston upon Thames	2.7	43.2	23.5	1.1	49.4	6.3	43.1	6.6	43.9	67	70.1
Merton	3.9	44.0	25.5	1.7	49.2	10.3	38.9	5.6	45.2	121	60.3
Redbridge	4.9	44.8	27.1	2.2	41.4	10.2	31.2	5.9	52.6	155	65.2
Richmond upon Thames	3.1	37.6	19.0	1.2	50.9	5.6	45.3	7.8	41.3	61	57.4
Sutton	3.0	51.6	23.5	1.5	58.4	7.5	51.0	5.4	36.2	108	72.2
Waltham Forest	5.6	54.1	29.9	3.0	44.6	10.3	34.3	5.4	50.0	191	53.4
South East	**118.6**	50.2	21.3	**59.6**	59.3	7.9	51.4	5.2	35.5	**5,088**	48.7
Bracknell Forest UA	**1.9**	47.4	21.8	**0.9**	55.8	6.0	49.7	6.5	37.7	**72**	55.6
Brighton and Hove UA	**4.4**	62.2	30.2	**2.7**	53.2	7.3	46.0	4.7	42.1	**184**	56.5
Isle of Wight UA	**1.5**	60.5	18.5	**0.9**	69.0	12.0	57.0	4.9	26.1	**81**	45.7
Medway UA	**4.0**	58.8	22.9	**2.4**	63.5	8.7	54.8	4.9	31.6	**218**	38.1
Milton Keynes UA	**4.1**	53.7	22.7	**2.2**	61.1	8.9	52.2	5.1	33.8	**181**	44.8
Portsmouth UA	**3.1**	62.4	25.8	**2.0**	58.1	10.5	47.7	4.4	37.4	**191**	45.5
Reading UA	**3.0**	58.5	30.0	**1.8**	48.2	8.8	39.4	4.0	47.9	**143**	46.9
Slough UA	**2.7**	44.1	24.8	**1.2**	53.1	11.2	41.9	5.6	41.3	**94**	40.4
Southampton UA	**3.6**	61.9	23.8	**2.2**	61.6	10.4	51.2	4.3	34.1	**204**	38.7
West Berkshire UA	**2.1**	46.3	17.5	**1.0**	59.1	6.6	52.5	5.9	35.1	**89**	48.3
Windsor and Maidenhead UA	**2.1**	42.2	22.2	**0.9**	50.1	6.3	43.8	7.6	42.3	**73**	58.9
Wokingham UA	**2.1**	38.6	18.5	**0.8**	50.8	5.3	45.6	5.4	43.8	**59**	74.6
Buckinghamshire	6.9	40.4	20.1	2.8	55.8	6.7	49.1	5.2	39.0	204	64.7
East Sussex	5.9	55.7	20.1	3.3	64.3	8.7	55.7	5.2	30.5	347	45.0
Hampshire	16.8	47.9	18.3	8.1	61.8	7.5	54.3	5.9	32.3	722	46.4
Kent	19.6	54.9	21.5	10.7	63.4	8.4	55.0	4.5	32.1	1,018	45.1
Oxfordshire	9.4	47.1	20.3	4.4	56.0	7.4	48.6	6.4	37.6	376	50.0
Surrey	15.4	42.4	21.0	6.5	54.5	6.0	48.5	5.1	40.4	427	61.8
West Sussex	10.0	48.6	19.9	4.9	61.4	7.2	54.2	5.7	32.8	405	49.1
South West	**65.4**	54.5	19.7	**35.7**	63.1	8.7	54.4	5.3	31.6	**3,234**	47.2
Bath and North East Somerset UA	**2.1**	52.5	21.9	**1.1**	58.0	7.5	50.4	4.9	37.1	**90**	56.7
Bournemouth UA	**2.4**	60.7	30.4	**1.5**	52.1	10.0	42.1	4.1	43.8	**94**	57.4
Bristol, City of UA	**6.9**	59.6	21.9	**4.1**	62.2	11.9	50.3	3.7	34.0	**314**	37.3
North Somerset UA	**2.5**	52.9	18.5	**1.3**	62.3	6.8	55.5	6.0	31.7	**90**	48.9
Plymouth UA	**3.5**	61.9	21.4	**2.2**	65.8	10.0	55.7	4.9	29.4	**225**	40.9
Poole UA	**1.8**	51.3	19.2	**0.9**	62.7	8.7	54.0	5.4	31.8	**71**	54.9
South Gloucestershire UA	**3.5**	47.9	16.4	**1.7**	64.0	7.6	56.5	5.0	31.0	**146**	43.2
Swindon UA	**3.0**	55.8	20.1	**1.7**	63.3	8.0	55.3	5.0	31.7	**198**	41.9
Torbay UA	**1.7**	61.8	22.3	**1.1**	62.9	9.2	53.7	5.4	31.7	**123**	48.8

1 Conceptions leading to births outside marriage registered by the mother alone.
2 Conceptions leading to births outside marriage registered by both parents.

Table 12.7 - *continued*

Area of usual residence	Total number of conceptions to all women (000s)	Percentage of all conceptions		Total number of conceptions outside marriage (000s)	Percentage of conceptions outside marriage					Total number of conceptions to women aged under 18	Percentage of under 18 conceptions terminated by abortion
		Outside marriage	Terminated by abortion		Leading to maternities				Terminated by abortion		
					Outside marriage			Within marriage following marriage after conception			
					Total	Sole[1]	Joint[2]				
Cornwall and the Isles of Scilly	5.9	58.0	18.7	3.4	66.1	8.2	57.9	5.9	28.1	324	45.1
Devon	8.1	53.0	17.8	4.3	65.8	7.8	58.0	5.4	28.8	405	46.4
Dorset	4.2	53.0	18.1	2.2	63.5	7.6	55.9	6.2	30.2	213	44.1
Gloucestershire	7.5	53.0	20.7	4.0	60.9	9.0	52.0	5.4	33.7	387	55.0
Somerset	6.3	54.9	18.2	3.4	66.3	8.2	58.2	5.1	28.5	322	49.4
Wiltshire	5.9	46.6	17.9	2.7	60.7	7.0	53.7	7.0	32.2	232	53.0
Wales	**40.3**	61.8	19.1	**24.9**	68.6	10.7	57.9	4.1	27.4	2,605	38.5
Isle of Anglesey	0.8	60.7	15.8	0.5	72.5	8.9	63.6	3.8	23.7	43	46.5
Gwynedd	1.5	65.0	18.8	1.0	71.1	8.3	62.8	2.4	26.5	79	36.7
Conwy	1.3	63.0	20.7	0.8	68.6	8.7	59.9	3.0	28.4	99	40.4
Denbighshire	1.2	62.9	19.5	0.8	67.9	8.7	59.2	4.8	27.3	92	38.0
Flintshire	2.1	57.3	19.1	1.2	65.3	9.1	56.1	5.4	29.3	107	45.8
Wrexham	2.0	62.8	21.9	1.2	65.0	10.4	54.6	4.4	30.6	146	39.0
Powys	1.5	55.1	14.9	0.8	72.0	7.7	64.3	4.7	23.3	75	49.3
Ceredigion	0.7	56.8	17.3	0.4	66.3	7.9	58.4	4.6	29.1	51	51.0
Pembrokeshire	1.4	62.2	19.3	0.9	67.7	9.5	58.3	4.2	28.0	96	38.5
Carmarthenshire	2.1	58.5	14.6	1.2	72.8	8.8	64.0	4.5	22.6	128	38.3
Swansea	2.9	61.5	16.3	1.8	71.8	12.7	59.1	4.4	23.8	173	24.3
Neath Port Talbot	1.8	64.2	16.6	1.2	71.2	11.1	60.1	4.9	23.9	118	29.7
Bridgend	1.8	60.9	17.9	1.1	70.0	10.6	59.4	4.1	25.9	118	32.2
The Vale of Glamorgan	1.6	56.7	18.6	0.9	67.7	11.5	56.2	4.4	27.9	97	45.4
Cardiff	5.0	57.7	22.1	2.9	63.4	12.0	51.4	3.5	33.1	273	40.7
Rhondda, Cynon, Taff	3.6	68.6	20.4	2.4	69.5	10.8	58.7	4.0	26.5	282	37.6
Merthyr Tydfil	0.8	72.7	16.7	0.6	75.6	13.3	62.4	3.7	20.7	77	28.6
Caerphilly	2.5	64.6	19.3	1.6	70.3	11.2	59.1	3.2	26.5	189	36.0
Blaenau Gwent	0.9	74.0	22.5	0.7	67.7	9.7	58.0	4.0	28.3	71	43.7
Torfaen	1.4	67.8	20.2	0.9	70.4	11.5	58.9	3.6	25.9	128	32.8
Monmouthshire	1.0	50.3	19.7	0.5	63.0	7.6	55.4	5.0	32.0	38	52.6
Newport	2.2	61.2	21.2	1.3	65.6	15.0	50.6	4.2	30.2	125	51.2

1 Conceptions leading to births outside marriage registered by the mother alone.
2 Conceptions leading to births outside marriage registered by both parents.

Table 12.8 Conceptions (numbers and percentages): occurrence within/outside marriage and outcome, area of usual residence, and age of woman at conception, 2004

England and Wales, Government Office Regions and health authorities (within England)
Residents

Area of usual residence	Total number of conceptions to all women (000s)	Percentage of all conceptions		Total number of conceptions outside marriage (000s)	Percentage of conceptions outside marriage					Total number of conceptions to women aged under 18	Percentage of under 18 conceptions terminated by abortion
		Outside marriage	Terminated by abortion		Leading to maternities				Terminated by abortion		
					Outside marriage			Within marriage following marriage after conception			
					Total	Sole[1]	Joint[2]				
England and Wales	826.8	54.9	22.4	454.1	60.4	10.0	50.5	4.6	35.0	42,198	45.6
England	786.5	54.6	22.6	429.2	59.9	9.9	50.0	4.6	35.5	39,593	46.0
North East	35.0	64.5	20.3	22.6	68.0	11.2	56.8	3.6	28.5	2,515	40.2
County Durham and Tees Valley	16.2	65.4	19.1	10.6	70.1	11.6	58.5	3.6	26.3	1,213	38.5
Northumberland, Tyne & Wear	18.8	63.8	21.3	12.0	66.1	10.8	55.2	3.6	30.4	1,302	41.7
North West	102.2	59.7	20.2	61.0	66.0	11.8	54.1	3.5	30.5	6,268	42.5
Cheshire & Merseyside	33.1	62.2	20.8	20.6	66.6	13.5	53.2	3.1	30.3	1,917	48.1
Cumbria and Lancashire	26.7	56.3	17.9	15.0	67.8	10.0	57.8	4.3	27.8	1,673	39.6
Greater Manchester	42.4	59.8	21.3	25.4	64.3	11.6	52.7	3.3	32.3	2,678	40.3
Yorkshire and the Humber	75.0	57.1	19.3	42.8	65.6	10.5	55.1	4.2	30.2	4,698	38.8
North and East Yorkshire and North Lincolnshire	21.6	58.5	18.8	12.6	66.5	9.5	57.1	4.8	28.6	1,400	41.0
South Yorkshire	19.1	61.9	20.9	11.8	66.2	10.3	55.9	3.7	30.1	1,351	39.7
West Yorkshire	34.4	53.5	18.7	18.4	64.7	11.4	53.3	4.1	31.3	1,947	36.7
East Midlands	60.5	55.8	19.7	33.8	65.3	9.7	55.6	4.6	30.1	3,368	40.9
Leicestershire, Northamptonshire and Rutland	24.6	52.4	20.0	12.9	63.6	9.1	54.5	4.6	31.8	1,211	43.6
Trent	35.9	58.2	19.5	20.9	66.4	10.0	56.3	4.6	29.0	2,157	39.5
West Midlands	84.4	54.7	22.0	46.1	62.1	10.9	51.2	4.1	33.8	4,795	43.2
Birmingham and the Black Country	41.6	53.6	23.4	22.3	60.4	12.7	47.7	3.6	36.1	2,436	42.8
Shropshire and Staffordshire	20.7	57.1	19.1	11.8	67.0	9.3	57.7	4.7	28.4	1,200	40.6
West Midlands South	22.1	54.3	22.3	12.0	60.6	9.3	51.4	4.6	34.8	1,159	46.8
East	79.5	50.9	19.2	40.5	62.3	8.3	54.0	5.3	32.3	3,392	47.0
Bedford and Hertfordshire	26.1	47.0	20.0	12.3	59.2	8.4	50.7	5.1	35.7	969	50.2
Essex	23.8	54.3	20.6	12.9	61.2	8.1	53.2	4.7	34.1	1,032	50.2
Norfolk, Suffolk and Cambridgeshire	29.6	51.6	17.5	15.3	65.8	8.5	57.3	6.0	28.1	1,391	42.3
London	165.9	52.6	30.8	87.2	45.7	10.1	35.6	5.0	49.3	6,235	59.1
North Central London	27.2	52.9	31.6	14.4	43.9	9.9	33.9	5.7	50.4	1,041	59.3
North East London	38.1	52.3	30.3	20.0	45.3	10.7	34.6	4.9	49.8	1,423	58.4
North West London	39.0	46.2	31.2	18.0	41.2	9.0	32.2	5.4	53.4	1,260	58.7
South East London	35.5	62.5	33.3	22.2	48.9	11.3	37.6	4.2	46.9	1,616	59.1
South West London	26.0	48.7	26.8	12.7	49.2	9.0	40.1	5.4	45.4	895	60.6

1 Conceptions leading to births outside marriage registered by the mother alone.
2 Conceptions leading to births outside marriage registered by both parents.

Table 12.8 - *continued*

Area of usual residence	Total number of conceptions to all women (000s)	Percentage of all conceptions		Total number of conceptions outside marriage (000s)	Percentage of conceptions outside marriage					Total number of conceptions to women aged under 18	Percentage of under 18 conceptions terminated by abortion
		Outside marriage	Terminated by abortion		Leading to maternities				Terminated by abortion		
					Outside marriage			Within marriage following marriage after conception			
					Total	Sole[1]	Joint[2]				
South East	**118.6**	50.2	21.3	**59.6**	59.3	7.9	51.4	5.2	35.5	**5,088**	48.7
Hampshire and Isle of Wight	**25.0**	52.5	20.0	**13.1**	61.7	8.7	53.0	5.4	32.9	**1,198**	44.9
Kent and Medway	**23.6**	55.5	21.8	**13.1**	63.4	8.5	54.9	4.5	32.0	**1,236**	43.9
Surrey and Sussex	**35.7**	48.8	21.7	**17.4**	58.1	7.0	51.1	5.2	36.7	**1,363**	53.0
Thames Valley	**34.4**	46.5	21.7	**16.0**	55.2	7.6	47.5	5.7	39.2	**1,291**	52.4
South West	**65.4**	54.5	19.7	**35.7**	63.1	8.7	54.4	5.3	31.6	**3,234**	47.2
Avon, Gloucestershire and Wiltshire	**31.4**	52.9	19.8	**16.6**	61.7	8.9	52.8	5.2	33.1	**1,457**	47.6
Dorset and Somerset	**14.7**	54.8	20.3	**8.1**	62.6	8.4	54.2	5.3	32.1	**700**	49.4
South West Peninsula	**19.3**	56.9	19.1	**11.0**	65.6	8.5	57.1	5.4	29.0	**1,077**	45.1
Wales	**40.3**	61.8	19.1	**24.9**	68.6	10.7	57.9	4.1	27.4	**2,605**	38.5

1 Conceptions leading to births outside marriage registered by the mother alone.
2 Conceptions leading to births outside marriage registered by both parents.

Table 12.9 Conceptions (numbers and rates): age of woman at conception, outcome, and area of usual residence, 2004

England and Wales, Government Office Regions (within England), counties, unitary authorities and London boroughs
Residents

Area of usual residence	All conceptions				Conceptions at ages under 18				Conceptions at ages under 16			
	Number (000s)	Rates per 1,000 women aged 15-44			Number	Rates per 1,000 women aged 15-17			Number	Rates per 1,000 women aged 13-15		
		Total	Maternities	Abortions		Total	Maternities	Abortions		Total	Maternities	Abortions
England and Wales	**826.8**	**75.3**	**58.4**	**16.8**	**42,198**	**41.7**	**22.7**	**19.0**	**7,615**	**7.5**	**3.2**	**4.3**
England	786.5	75.6	58.6	17.0	39,593	41.5	22.4	19.1	7,181	7.5	3.2	4.3
North East	**35.0**	**67.5**	**53.8**	**13.7**	2,515	50.6	30.3	20.3	481	9.8	4.7	5.1
Darlington UA	**1.5**	**76.4**	**63.2**	**13.2**	93	47.8	27.7	20.0	19	*9.6*	*	*
Hartlepool UA	**1.4**	**76.8**	**59.6**	**17.2**	126	64.1	37.1	27.0	21	10.9	*4.7*	*6.2*
Middlesbrough UA	**2.4**	**80.5**	**64.2**	**16.3**	191	61.6	38.4	23.2	36	12.3	*4.4*	7.8
Redcar and Cleveland UA	**1.8**	**67.8**	**54.5**	**13.3**	173	58.3	38.8	19.6	54	18.4	10.6	7.8
Stockton-on-Tees UA	**2.8**	**70.6**	**56.3**	**14.3**	185	49.0	30.0	19.1	39	10.3	5.5	*4.7*
Durham	**6.3**	63.8	52.5	11.3	445	48.2	29.5	18.7	93	10.0	4.6	5.4
Northumberland	**3.6**	63.9	51.3	12.6	214	37.4	20.6	16.8	40	7.0	*3.3*	3.7
Tyne and Wear	**15.2**	66.2	51.8	14.3	1,088	51.8	30.5	21.3	179	8.8	4.0	4.8
North West	**102.2**	**72.6**	**57.9**	**14.7**	6,268	45.6	26.2	19.4	1,076	7.8	3.6	4.3
Blackburn with Darwin UA	**2.8**	**93.2**	**78.7**	**14.5**	179	52.1	32.0	20.1	28	8.5	*2.7*	*5.8*
Blackpool UA	**2.1**	**77.9**	**60.9**	**17.0**	193	72.3	49.4	22.8	33	12.3	*	*
Halton UA	**2.0**	**78.8**	**62.1**	**16.7**	115	43.8	25.9	17.9	10	*3.9*	*	*
Warrington UA	**2.8**	**69.0**	**55.0**	**14.0**	153	40.6	23.1	17.5	20	5.1	*2.3*	2.8
Cheshire	**9.0**	67.7	55.7	12.0	425	33.4	17.2	16.2	74	5.7	2.2	3.5
Cumbria	**5.9**	64.2	52.6	11.6	333	35.9	20.5	15.4	59	6.3	2.9	3.4
Greater Manchester	**42.4**	77.8	61.2	16.5	2,678	52.4	31.3	21.1	487	9.5	4.4	5.1
Lancashire	**15.9**	69.3	57.0	12.3	968	41.9	25.1	16.8	159	6.8	3.2	3.6
Merseyside	**19.4**	67.7	52.6	15.1	1,224	42.6	21.6	21.0	206	7.3	3.1	4.2
Yorkshire and the Humber	**75.0**	**72.4**	**58.4**	**13.9**	4,698	47.3	28.9	18.4	871	8.7	4.1	4.6
East Riding of Yorkshire UA	**3.5**	**60.9**	**50.6**	**10.3**	190	31.7	17.2	14.5	50	8.1	4.6	3.6
Kingston upon Hull, City of UA	**4.1**	**75.9**	**60.3**	**15.6**	412	79.0	53.5	25.5	77	15.1	8.0	7.1
North East Lincolnshire UA	**2.5**	**79.7**	**62.4**	**17.3**	239	65.9	39.4	26.5	47	13.3	*6.5*	6.8
North Lincolnshire UA	**2.2**	**73.0**	**59.4**	**13.6**	156	51.5	29.0	22.4	26	8.3	*2.9*	*5.4*
York UA	**2.4**	**58.4**	**46.2**	**12.2**	113	35.1	18.4	16.8	15	*4.8*	*1.3*	*3.5*
North Yorkshire	**6.9**	65.7	54.6	11.1	290	25.5	13.6	12.0	42	3.6	*1.0*	2.6
South Yorkshire	**19.1**	72.2	57.1	15.1	1,351	55.3	33.4	21.9	262	10.7	4.9	5.7
West Yorkshire	**34.4**	75.8	61.6	14.1	1,947	45.9	29.1	16.8	352	8.3	4.0	4.3
East Midlands	**60.5**	**69.9**	**56.2**	**13.8**	3,368	41.0	24.2	16.8	610	7.3	3.6	3.7
Derby UA	**3.8**	**76.4**	**59.5**	**16.9**	241	54.7	37.0	17.7	45	9.7	5.4	4.3
Leicester UA & Rutland UA	**6.4**	**85.7**	**66.1**	**19.5**	300	43.9	28.3	16.6	48	7.5	3.9	3.6
Nottingham UA	**4.9**	**72.6**	**53.5**	**19.1**	376	72.8	49.7	23.0	56	11.1	6.8	4.4
Derbyshire	**9.4**	65.9	55.0	10.9	506	36.6	20.1	16.5	99	6.9	3.3	3.5
Leicestershire	**8.0**	65.3	53.5	11.8	353	30.3	14.8	15.5	71	5.9	2.4	3.5
Lincolnshire	**8.0**	64.7	52.8	11.8	503	39.3	22.6	16.7	83	6.3	3.1	3.3
Northamptonshire	**10.2**	77.3	62.0	15.3	558	43.1	24.6	18.6	102	7.7	3.9	3.9
Nottinghamshire	**9.7**	64.4	52.3	12.1	531	36.5	21.9	14.6	106	7.2	3.3	3.9
West Midlands	**84.4**	**78.2**	**61.0**	**17.2**	4,795	45.0	25.6	19.5	899	8.4	3.7	4.7
Herefordshire, County of UA	**2.0**	**64.1**	**53.0**	**11.2**	126	37.8	20.4	17.4	23	6.5	*2.5*	*4.0*
Stoke-on-Trent UA	**4.1**	**82.1**	**67.5**	**14.6**	320	67.6	45.6	22.0	75	15.7	8.6	7.1
Telford and Wrekin UA	**2.6**	**75.6**	**60.4**	**15.2**	182	52.0	38.0	14.0	40	11.6	6.1	*5.5*
Shropshire	**3.4**	67.1	55.7	11.4	148	27.4	13.9	13.5	17	*3.1*	*	*
Staffordshire	**10.6**	67.7	54.2	13.5	550	34.6	18.2	16.4	119	7.4	3.5	3.9
Warwickshire	**7.1**	68.2	53.4	14.8	371	37.1	19.4	17.7	75	7.5	2.7	4.8
West Midlands	**47.2**	86.4	65.5	20.9	2,752	51.8	29.3	22.5	493	9.3	4.0	5.4
Worcestershire	**7.4**	69.8	57.2	12.6	346	33.0	18.5	14.5	57	5.3	2.2	3.2

Note: See sections 2.1 and 2.7.

Table 12.9 - *continued*

Area of usual residence	All conceptions				Conceptions at ages under 18				Conceptions at ages under 16			
	Number (000s)	Rates per 1,000 women aged 15-44			Number	Rates per 1,000 women aged 15-17			Number	Rates per 1,000 women aged 13-15		
		Total	Maternities	Abortions		Total	Maternities	Abortions		Total	Maternities	Abortions
East	**79.5**	73.0	59.0	14.1	3,392	32.8	17.4	15.4	600	5.7	2.4	3.3
Luton UA	**4.2**	102.8	79.2	23.5	182	45.2	23.9	21.4	31	8.3	*3.5*	*4.8*
Peterborough UA	**3.0**	88.8	71.7	17.0	175	53.8	37.2	16.6	32	9.8	6.1	*3.7*
Southend-on-Sea UA	**2.5**	80.9	62.5	18.4	135	47.4	27.7	19.7	27	9.3	*3.1*	*6.2*
Thurrock UA	**2.8**	87.6	68.9	18.7	123	43.1	22.4	20.7	14	*4.7*	*	*
Bedfordshire	**6.0**	74.1	59.5	14.6	242	32.3	15.6	16.7	31	4.1	*1.9*	*2.3*
Cambridgeshire	**7.6**	62.7	52.5	10.2	267	25.5	14.0	11.4	53	5.1	2.2	2.9
Essex	**18.4**	71.7	57.2	14.5	774	30.6	14.7	16.0	143	5.6	2.1	3.5
Hertfordshire	**15.9**	73.7	59.5	14.2	545	27.3	13.5	13.8	115	5.5	2.1	3.4
Norfolk	**10.0**	66.9	54.4	12.5	514	35.9	20.5	15.4	81	5.6	2.7	2.9
Suffolk	**9.0**	71.6	59.6	12.0	435	34.0	18.8	15.2	73	5.5	2.3	3.2
London	**165.9**	91.2	63.1	28.1	6,235	48.3	19.8	28.5	1,128	9.0	3.1	5.9
Inner London	**73.7**	92.5	61.3	31.2	2,705	58.3	23.7	34.6	485	11.2	4.0	7.2
Camden	**4.4**	70.8	48.6	22.2	127	42.1	14.3	27.9	14	*5.2*	*0.4*	*4.8*
Hackney & City of London	**6.6**	115.1	77.3	37.8	285	69.9	28.4	41.4	52	13.2	*3.0*	10.1
Hammersmith and Fulham	**3.9**	78.6	53.1	25.4	106	44.3	20.9	23.4	11	*4.9*	*	*
Haringey	**5.9**	101.5	66.9	34.7	283	68.8	32.1	36.7	62	16.4	7.4	9.0
Islington	**4.3**	84.3	53.7	30.6	156	54.8	17.9	36.9	35	13.4	*3.8*	9.6
Kensington and Chelsea	**3.1**	62.9	42.8	20.0	54	24.1	8.9	15.2	5	*2.8*	*	*
Lambeth	**7.5**	104.5	65.0	39.5	357	85.2	32.5	52.8	64	16.1	*4.8*	11.3
Lewisham	**6.5**	104.0	66.3	37.7	317	70.4	30.0	40.4	59	13.7	*4.2*	9.5
Newham	**7.7**	125.3	85.6	39.7	256	48.8	19.4	29.3	47	9.2	4.3	4.9
Southwark	**7.5**	113.6	70.1	43.4	343	85.5	33.4	52.1	66	16.5	6.7	9.7
Tower Hamlets	**5.6**	98.0	68.6	29.4	173	42.9	18.1	24.8	35	9.4	*3.2*	6.2
Wandsworth	**6.3**	75.3	54.1	21.3	174	58.0	26.6	31.3	27	9.2	*5.8*	*3.4*
Westminster	**4.3**	65.4	43.0	22.5	74	27.1	10.2	16.8	8	*3.7*	*	*
Outer London	**92.2**	90.1	64.4	25.7	3,530	42.7	17.6	25.2	643	7.9	2.6	5.3
Barking and Dagenham	**4.2**	112.7	78.8	34.0	211	72.5	33.7	38.8	42	13.4	*5.1*	8.3
Barnet	**6.5**	85.7	61.5	24.2	203	34.9	11.0	23.9	34	6.1	*1.8*	4.3
Bexley	**3.6**	79.1	58.6	20.6	172	38.3	13.1	25.1	36	7.8	*2.0*	5.9
Brent	**6.9**	106.0	68.7	37.3	266	53.4	22.3	31.1	49	10.5	*3.6*	6.9
Bromley	**4.7**	75.8	57.8	17.9	166	30.9	12.3	18.6	36	6.4	*2.1*	4.2
Croydon	**7.1**	90.6	61.3	29.3	364	55.4	22.7	32.7	63	9.7	3.1	6.6
Ealing	**6.8**	94.1	66.9	27.2	193	36.7	15.6	21.1	32	6.5	*1.8*	4.7
Enfield	**6.2**	98.3	69.3	28.9	272	51.3	25.3	26.0	46	8.9	3.1	5.8
Greenwich	**5.6**	101.8	70.9	30.9	261	64.7	32.4	32.2	49	12.3	5.3	7.0
Harrow	**4.1**	88.3	62.1	26.2	141	33.6	10.9	22.6	25	6.2	*1.5*	*4.7*
Havering	**3.4**	76.6	55.4	21.2	152	35.3	13.9	21.4	25	5.7	*1.8*	*3.9*
Hillingdon	**4.9**	85.3	60.5	24.8	223	46.4	17.7	28.7	50	10.4	*2.5*	7.9
Hounslow	**5.0**	100.6	72.5	28.0	203	52.2	25.5	26.8	36	9.8	*4.3*	5.4
Kingston upon Thames	**2.7**	73.9	56.5	17.4	67	25.6	7.6	18.0	9	*3.5*	*	*
Merton	**3.9**	83.9	62.5	21.4	121	39.4	15.6	23.8	24	8.0	*2.3*	*5.7*
Redbridge	**4.9**	90.7	66.1	24.6	155	32.4	11.3	21.1	25	5.1	*1.8*	*3.3*
Richmond upon Thames	**3.1**	73.6	59.6	14.0	61	22.2	9.5	12.7	12	*4.5*	*	*
Sutton	**3.0**	77.6	59.4	18.3	108	31.5	8.8	22.8	20	5.7	*2.0*	*3.7*
Waltham Forest	**5.6**	104.6	73.3	31.3	191	47.3	22.0	25.2	30	7.5	2.2	5.2
South East	**118.6**	72.7	57.2	15.5	5,088	33.5	17.2	16.3	918	5.9	2.2	3.7
Bracknell Forest UA	**1.9**	74.1	57.9	16.2	72	31.5	14.0	17.5	8	*3.5*	*	*
Brighton and Hove UA	**4.4**	71.7	50.1	21.7	184	44.4	19.3	25.1	38	9.5	*3.0*	6.5
Isle of Wight UA	**1.5**	62.5	50.9	11.6	81	32.1	17.4	14.7	12	4.6	*0.4*	4.2
Medway UA	**4.0**	74.8	57.7	17.1	218	40.7	25.2	15.5	43	7.9	3.7	4.2
Milton Keynes UA	**4.1**	87.0	67.2	19.8	181	41.9	23.2	18.8	36	8.1	*4.1*	*4.1*

Note: See sections 2.1 and 2.7.

Table 12.9 - *continued*

Area of usual residence	All conceptions				Conceptions at ages under 18				Conceptions at ages under 16			
	Number (000s)	Rates per 1,000 women aged 15-44			Number	Rates per 1,000 women aged 15-17			Number	Rates per 1,000 women aged 13-15		
		Total	Maternities	Abortions		Total	Maternities	Abortions		Total	Maternities	Abortions
Portsmouth UA	3.1	71.8	53.3	18.5	191	54.2	29.5	24.7	30	8.8	2.1	6.8
Reading UA	3.0	86.9	60.8	26.1	143	59.4	31.6	27.8	32	13.3	5.4	7.9
Slough UA	2.7	99.5	74.8	24.7	94	41.2	24.6	16.7	11	4.9	*	*
Southampton UA	3.6	68.5	52.2	16.3	204	56.1	34.4	21.7	36	9.9	4.7	5.2
West Berkshire UA	2.1	73.5	60.6	12.9	89	28.1	14.5	13.6	16	4.9	*	*
Windsor and Maidenhead UA	2.1	75.1	58.5	16.7	73	29.3	12.0	17.2	8	3.0	*	*
Wokingham UA	2.1	64.7	52.8	11.9	59	19.6	5.0	14.6	15	5.0	*	*
Buckinghamshire	6.9	72.9	58.3	14.6	204	21.9	7.7	14.2	27	2.8	0.5	2.3
East Sussex	5.9	70.2	56.1	14.1	347	37.5	20.6	16.9	60	6.4	2.9	3.5
Hampshire	16.8	68.9	56.3	12.6	722	30.3	16.2	14.1	125	5.2	2.0	3.2
Kent	19.6	74.1	58.1	15.9	1,018	38.1	20.9	17.2	184	6.7	2.9	3.8
Oxfordshire	9.4	70.8	56.4	14.4	376	34.3	17.1	17.1	73	6.4	2.1	4.3
Surrey	15.4	71.7	56.6	15.1	427	22.2	8.5	13.7	81	4.1	1.3	2.8
West Sussex	10.0	71.6	57.3	14.2	405	30.0	15.2	14.7	83	5.9	2.3	3.6
South West	65.4	68.4	54.9	13.5	3,234	34.5	18.2	16.3	598	6.3	2.5	3.8
Bath and North East Somerset UA	2.1	60.3	47.1	13.2	90	27.8	12.1	15.8	16	5.1	1.9	3.2
Bournemouth UA	2.4	70.1	48.8	21.3	94	37.1	15.8	21.3	25	9.5	4.9	4.6
Bristol, City of UA	6.9	73.1	57.1	16.0	314	46.7	29.3	17.4	59	9.1	4.5	4.6
North Somerset UA	2.5	73.6	60.0	13.6	90	25.5	13.0	12.4	16	4.5	1.7	2.8
Plymouth UA	3.5	68.9	54.2	14.7	225	47.2	27.9	19.3	38	8.4	3.1	5.3
Poole UA	1.8	71.4	57.7	13.7	71	26.3	11.8	14.4	11	4.2	*	*
South Gloucestershire UA	3.5	70.2	58.7	11.5	146	31.6	17.9	13.6	24	4.9	2.7	2.3
Swindon UA	3.0	77.1	61.6	15.5	198	56.4	32.8	23.6	26	7.6	3.2	4.4
Torbay UA	1.7	74.7	58.1	16.6	123	49.9	25.6	24.4	32	13.4	5.9	7.5
Cornwall and the Isles of Scilly	5.9	64.9	52.8	12.1	324	34.3	18.9	15.5	65	6.9	2.2	4.7
Devon	8.1	64.0	52.6	11.4	405	31.2	16.7	14.5	69	5.2	2.4	2.8
Dorset	4.2	65.7	53.8	11.9	213	28.7	16.0	12.7	45	5.9	3.0	2.9
Gloucestershire	7.5	67.6	53.6	14.0	387	34.4	15.4	18.9	81	7.2	2.0	5.2
Somerset	6.3	68.3	55.9	12.4	322	32.0	16.2	15.8	58	5.6	1.9	3.7
Wiltshire	5.9	69.3	56.9	12.4	232	26.9	12.7	14.3	33	3.7	0.9	2.8
Wales	40.3	69.5	56.2	13.3	2,605	45.1	27.7	17.3	434	7.5	3.8	3.7
Isle of Anglesey	0.8	64.5	54.3	10.2	43	32.5	17.4	15.1	7	5.2	*	*
Gwynedd	1.5	68.2	55.4	12.8	79	38.0	24.0	13.9	10	4.7	*	*
Conwy	1.3	70.4	55.8	14.6	99	49.4	29.4	20.0	17	8.2	*	*
Denbighshire	1.2	72.0	58.0	14.1	92	50.6	31.3	19.2	14	7.5	*	*
Flintshire	2.1	69.4	56.2	13.2	107	37.5	20.3	17.2	21	7.1	2.7	4.4
Wrexham	2.0	76.5	59.7	16.8	146	62.2	37.9	24.3	30	12.2	6.1	6.1
Powys	1.5	69.0	58.7	10.3	75	31.0	15.7	15.3	14	5.7	*	*
Ceredigion	0.7	47.0	38.8	8.2	51	40.5	19.8	20.6	8	6.5	*	*
Pembrokeshire	1.4	70.3	56.8	13.6	96	42.0	25.8	16.2	12	5.0	*	*
Carmarthenshire	2.1	63.8	54.5	9.3	128	36.7	22.7	14.1	14	4.2	*	*
Swansea	2.9	65.5	54.8	10.7	173	41.2	31.2	10.0	21	5.1	2.7	2.4
Neath Port Talbot	1.8	70.2	58.5	11.6	118	44.2	31.1	13.1	17	6.4	*	*
Bridgend	1.8	72.0	59.1	12.9	118	44.2	30.0	14.2	19	7.2	*	*
The Vale of Glamorgan	1.6	67.3	54.8	12.5	97	36.9	20.2	16.7	21	8.3	2.4	5.9
Cardiff	5.0	65.7	51.2	14.5	273	45.0	26.7	18.3	44	7.4	3.9	3.5
Rhondda, Cynon, Taff	3.6	75.6	60.2	15.4	282	59.4	37.1	22.3	62	13.3	6.6	6.6
Merthyr Tydfil	0.8	74.0	61.6	12.4	77	64.7	46.2	18.5	9	7.6	*	*
Caerphilly	2.5	73.7	59.5	14.2	189	53.1	34.0	19.1	31	8.9	6.0	2.9
Blaenau Gwent	0.9	67.6	52.4	15.2	71	46.0	25.9	20.1	12	7.9	*	*
Torfaen	1.4	78.9	63.0	15.9	128	66.9	45.0	22.0	21	11.0	5.2	5.7
Monmouthshire	1.0	68.1	54.6	13.4	38	21.6	10.2	11.4	9	5.1	*	*
Newport	2.2	77.4	61.0	16.4	125	41.9	20.4	21.4	21	7.0	3.0	4.0

Note: See sections 2.1 and 2.7.

Table 12.10 Conceptions (numbers and rates): age of woman at conception, outcome, and area of usual residence, 2004

England and Wales, Government Office
Regions, and health authorities
(within England)
Residents

Area of usual residence	All conceptions				Conceptions at ages under 18				Conceptions at ages under 16			
	Number (000s)	Rates per 1,000 women aged 15-44			Number	Rates per 1,000 women aged 15-17		women	Number	Rates per 1,000 women aged 13-15		
		Total	Maternities	Abortions		Total	Maternities	Abortions		Total	Maternities	Abortions
England and Wales	826.8	75.3	58.4	16.8	42,198	41.7	22.7	19.0	7,615	7.5	3.2	4.3
England	786.5	75.6	58.6	17.0	39,593	41.5	22.4	19.1	7,181	7.5	3.2	4.3
North East	35.0	67.5	53.8	13.7	2,515	50.6	30.3	20.3	481	9.8	4.7	5.1
County Durham and Tees Valley	16.2	69.6	56.3	13.3	1,213	52.8	32.5	20.3	262	11.4	5.6	5.8
Northumberland, Tyne & Wear	18.8	65.7	51.7	14.0	1,302	48.7	28.4	20.3	219	8.4	3.9	4.5
North West	102.2	72.6	57.9	14.7	6,268	45.6	26.2	19.4	1,076	7.8	3.6	4.3
Cheshire & Merseyside	33.1	68.4	54.2	14.3	1,917	40.1	20.8	19.3	310	6.5	2.7	3.8
Cumbria and Lancashire	26.7	70.6	58.0	12.6	1,673	43.5	26.3	17.2	279	7.2	3.5	3.7
Greater Manchester	42.4	77.8	61.2	16.5	2,678	52.4	31.3	21.1	487	9.5	4.4	5.1
Yorkshire and the Humber	75.0	72.4	58.4	13.9	4,698	47.3	28.9	18.4	871	8.7	4.1	4.6
North and East Yorkshire and Northern Lincolnshire	21.6	67.6	54.9	12.7	1,400	43.2	25.5	17.7	257	7.9	3.6	4.3
South Yorkshire	19.1	72.2	57.1	15.1	1,351	55.3	33.4	21.9	262	10.7	4.9	5.7
West Yorkshire	34.4	75.8	61.6	14.1	1,947	45.9	29.1	16.8	352	8.3	4.0	4.3
East Midlands	60.5	69.9	56.2	13.8	3,368	41.0	24.2	16.8	610	7.3	3.6	3.7
Leicestershire, Northamptonshire and Rutland	24.6	74.7	59.8	14.9	1,211	38.6	21.8	16.8	221	7.0	3.3	3.7
Trent	35.9	67.0	53.9	13.1	2,157	42.5	25.7	16.8	389	7.5	3.8	3.7
West Midlands	84.4	78.2	61.0	17.2	4,795	45.0	25.6	19.5	899	8.4	3.7	4.7
Birmingham and the Black Country	41.6	86.5	66.3	20.2	2,436	51.7	29.6	22.1	441	9.4	4.1	5.3
Shropshire and Staffordshire	20.7	71.0	57.5	13.5	1,200	40.6	24.1	16.5	251	8.4	4.2	4.2
West Midlands South	22.1	72.0	55.9	16.1	1,159	38.9	20.7	18.2	207	6.9	2.5	4.4
East	79.5	73.0	59.0	14.1	3,392	32.8	17.4	15.4	600	5.7	2.4	3.3
Bedfordshire and Hertfordshire	26.1	77.3	61.9	15.4	969	30.8	15.4	15.4	177	5.5	2.2	3.3
Essex	23.8	74.2	58.9	15.3	1,032	33.3	16.6	16.7	184	5.9	2.2	3.7
Norfolk, Suffolk and Cambridgeshire	29.6	68.8	56.8	12.1	1,391	34.1	19.6	14.4	239	5.8	2.7	3.1
London	165.9	91.2	63.1	28.1	6,235	48.3	19.8	28.5	1,128	9.0	3.1	5.9
North Central London	27.2	88.0	60.3	27.8	1,041	49.4	20.1	29.3	191	9.6	3.3	6.4
North East London	38.1	104.1	72.6	31.5	1,423	48.4	20.1	28.3	256	8.8	3.0	5.7
North West London	39.0	85.6	58.9	26.7	1,260	41.3	17.1	24.2	216	7.6	2.4	5.3
South East London	35.5	97.6	65.0	32.5	1,616	60.7	24.8	35.9	310	11.7	4.0	7.7
South West London	26.0	80.1	58.6	21.5	895	41.8	16.5	25.3	155	7.3	2.7	4.6
South East	118.6	72.7	57.2	15.5	5,088	33.5	17.2	16.3	918	5.9	2.2	3.7
Hampshire and Isle of Wight	25.0	68.8	55.0	13.8	1,198	35.7	19.7	16.1	203	6.0	2.2	3.9

Note: See section 2.1.

Table 12.10 - *continued*

Area of usual residence	All conceptions				Conceptions at ages under 18				Conceptions at ages under 16			
	Number (000s)	Rates per 1,000 women aged 15-44			Number	Rates per 1,000 women aged 15-17			Number	Rates per 1,000 women aged 13-15		
		Total	Maternities	Abortions		Total	Maternities	Abortions		Total	Maternities	Abortions
Kent and Medway	23.6	74.2	58.0	16.1	1,236	38.6	21.7	16.9	227	6.9	3.0	3.9
Surrey and Sussex	35.7	71.4	55.9	15.5	1,363	29.5	13.9	15.7	262	5.6	2.1	3.5
Thames Valley	34.4	76.1	59.6	16.5	1,291	32.1	15.3	16.8	226	5.5	1.9	3.6
South West	**65.4**	68.4	54.9	13.5	3,234	34.5	18.2	16.3	598	6.3	2.5	3.8
Avon, Gloucestershire and Wiltshire	31.4	70.1	56.2	13.9	1,457	35.1	18.4	16.7	255	6.1	2.3	3.8
Dorset and Somerset	14.7	68.2	54.4	13.8	700	30.8	15.6	15.2	139	6.0	2.5	3.4
South West Peninsula	19.3	66.0	53.4	12.6	1,077	36.3	19.9	16.4	204	6.9	2.7	4.1
Wales	**40.3**	69.5	56.2	13.3	2,605	45.1	27.7	17.3	434	7.5	3.8	3.7

Note: See section 2.1.

Appendix Table 1   Estimated resident population[1]: sex and age, 1994-2004

England and Wales

thousands

| | Age | Year | | | | | | | | | | |
		1994	1995	1996	1997	1998	1999	2000	2001	2002	2003	2004
Persons	**All ages**	**51,116.2**	**51,272.0**	**51,410.4**	**51,559.6**	**51,720.1**	**51,933.5**	**52,140.2**	**52,360.0**	**52,570.2**	**52,793.7**	**53,045.6**
Males	All ages	24,853.0	24,946.3	25,029.8	25,113.2	25,200.7	25,323.4	25,438.0	25,574.3	25,702.4	25,840.6	25,988.2
Females	All ages	26,263.2	26,325.6	26,380.7	26,446.5	26,519.4	26,610.0	26,702.1	26,785.7	26,867.8	26,953.1	27,057.4
	15-44	**10,720.9**	**10,711.9**	**10,717.6**	**10,727.0**	**10,737.2**	**10,766.6**	**10,820.8**	**10,869.5**	**10,905.2**	**10,944.1**	**10,983.5**
	15-19	1,454.1	1,473.8	1,501.4	1,534.3	1,560.7	1,566.5	1,562.7	1,578.4	1,609.3	1,649.6	1,678.7
	20-24	1,774.3	1,712.0	1,633.0	1,560.6	1,516.6	1,516.9	1,540.0	1,577.7	1,604.4	1,636.8	1,665.3
	25-29	2,041.3	2,004.9	1,979.7	1,945.2	1,902.3	1,850.9	1,809.9	1,744.7	1,674.0	1,627.5	1,625.4
	30-34	2,008.0	2,051.0	2,075.8	2,087.9	2,080.7	2,069.2	2,049.3	2,033.4	2,010.3	1,974.5	1,916.4
	35-39	1,760.0	1,802.5	1,853.9	1,903.4	1,951.9	2,004.0	2,053.7	2,082.0	2,102.5	2,101.0	2,092.5
	40-44	1,683.3	1,667.7	1,673.8	1,695.6	1,725.0	1,759.1	1,805.2	1,853.3	1,904.8	1,954.6	2,005.1

Note: Figures may not add exactly due to rounding - see section 2.7.
1 See section 2.1.

Appendix Table 2   Estimated resident female population[1]: age and marital status, 1994–2004

England and Wales

thousands

| | Age | Year | | | | | | | | | | |
		1994	1995	1996	1997	1998	1999	2000	2001	2002	2003	2004
Married	**15-44**	**5,309.2**	**5,178.1**	**5,070.5**	**4,962.6**	**4,859.5**	**4,772.1**	**4,696.7**	**4,611.1**	**4,490.8**	**4,372.4**	**4,262.0**
	15-19	21.6	20.8	20.7	20.3	20.3	19.8	17.6	16.2	13.0	11.7	10.6
	20-24	345.0	300.0	259.6	224.9	201.1	187.8	179.6	178.0	166.1	160.8	156.4
	25-29	1,022.3	962.7	906.3	844.3	782.8	725.1	677.2	625.4	567.4	523.5	496.6
	30-34	1,336.4	1,330.4	1,315.8	1,293.1	1,259.4	1,222.8	1,182.2	1,142.4	1,094.5	1,042.6	984.6
	35-39	1,297.1	1,305.5	1,319.9	1,331.5	1,342.0	1,354.1	1,362.1	1,356.3	1,342.4	1,314.6	1,284.4
	40-44	1,287.0	1,258.7	1,248.1	1,248.4	1,253.9	1,262.5	1,278.0	1,292.9	1,307.4	1,319.1	1,329.4
Single, widowed and divorced	**15-44**	**5,411.7**	**5,533.9**	**5,647.1**	**5,764.4**	**5,877.7**	**5,994.6**	**6,124.1**	**6,258.4**	**6,414.4**	**6,571.7**	**6,721.4**
	15-19	1,432.5	1,453.1	1,480.7	1,514.0	1,540.4	1,546.8	1,545.1	1,562.3	1,596.3	1,637.9	1,668.1
	20-24	1,429.3	1,412.0	1,373.3	1,335.7	1,315.6	1,329.1	1,360.5	1,399.7	1,438.3	1,476.0	1,508.9
	25-29	1,019.0	1,042.2	1,073.5	1,100.8	1,119.4	1,125.8	1,132.7	1,119.3	1,106.6	1,104.0	1,128.8
	30-34	671.6	720.6	760.0	794.7	821.3	846.4	867.1	891.0	915.8	931.8	931.8
	35-39	463.0	497.0	534.0	571.9	609.9	649.8	691.6	725.7	760.0	786.5	808.1
	40-44	396.4	409.0	425.7	447.3	471.1	496.7	527.3	560.5	597.4	635.5	675.7

Note: Figures may not add exactly due to rounding - see section 2.7.
1 See section 2.1.